High Interest/Low Readability

Writing

Grade 6

by

Tanya Anderson

Published by Milestone
an imprint of
Frank Schaffer Publications®

Author: Tanya Anderson
Editor: Krista Fanning
Interior Designer: Lori Kibbey

Frank Schaffer Publications®

Milestone is an imprint of Frank Schaffer Publications.

Send all inquiries to:
Frank Schaffer Publications
3195 Wilson Drive NW
Grand Rapids, Michigan 49534

High Interest/Low Readability Writing—grade 6

ISBN: 0-7696-4026-5

3 4 5 6 7 8 9 10 PAT 09 08 07

Table of Contents

To the Teacher:

Welcome to *High Interest/Low Readability Writing* for grade 6. This series was designed with the following purposes in mind:

- to help you, the teacher, meet your objectives and state standards
- to give you tools to motivate your students toward competence
- to provide extra practice for your students in key writing areas
- to provide enrichment activities to challenge your students
- to give you additional assessment tools for students' writing

In this series, you will find high-interest pages that not only challenge your students, but also motivate them to want to write. The student-friendly topics and clear directions will allow most students to work independently. To encourage fluency and skill development, all the directions and reading passages are written in the third- to sixth-grade reading level. When students read text at their appropriate instructional level, they gain confidence and experience success.

In this book, there are several recurring elements.

The Real Deal gives examples of the writing being introduced, which serve as models for students' own efforts.

Words to the Wise reinforces concepts and skills that support and enhance the writing unit.

Tools to Use includes graphic organizers, reference materials, technology aids, and other organizational, formatting, and research tools.

Focus on Form provides activities that introduce and reinforce writing elements and genres.

Your Turn! initiates the student's writing for that unit.

Revise for the Prize focuses on revision and proofreading to create a polished final draft. Final drafts are then placed in the student's Personal Portfolio.

Read and Write invites students to respond to authentic literature.

Write to Know Trivia are high-interest facts that show students that writing is both fun and informational.

High Interest/Low Readability Writing also provides simple assessment tools for writing a paragraph, a letter, and an essay. These writing exercises can be used by you, by other students for peer assessment, or by the student for self-assessment.

As students learn to write, they also discover that they write to learn. Reading, research, inquiry, and investigation are key parts of the writing journey. *High Interest/Low Readability Writing* is a tool that will help them find their way.

To the Writer:

Yes, that's you—the writer! All you need is a pencil, paper, and an idea, and off you go—becoming a writer! This book will give you many of the tools you need to grow as a writer.

In many ways, learning to write is like learning a sport.

- There are rules.
- The rules are there for good reasons.
- The best athletes follow the rules.
- The best athletes practice, practice, practice.
- The best athletes make mistakes, but they learn how to improve so they don't make the same mistakes again.
- The best athletes keep learning about the game.
- The best athletes know that they can get help from others when they need it.
- Athletes play in front of an audience.
- The best athletes are ready to play their best for the big game.

Learning to write is very similar.

✓ There are rules, such as correct grammar, punctuation, formatting, etc.
✓ The rules are there for good reasons. They make your writing clear for your reader.
✓ Good writers always follow the rules.
✓ Good writers practice. They write and write and write.
✓ All writers make mistakes, but they learn how to revise and correct them.
✓ Good writers keep learning about how to improve their skills.
✓ Good writers know how to use resources, such as a dictionary, a thesaurus, and other tools.
✓ Good writers write for their audience: the reader.
✓ Good writers polish their work to create an excellent final draft.

This book will help you as you practice becoming a good writer. It was designed to help you have a little fun along the way. Practice good writing, and you will be a writing winner!

Correlations	Word Analysis and Vocabulary Development	Literary Response and Analysis	Writing Strategies	Writing Application	Written and Oral English Language Conventions
Sentences	10, 12		7, 9		7, 8, 9, 10
Paragraphs—Basic			11, 13	11, 14, 75	
Graphic Organizers			12, 18, 41, 64, 66, 69		
Expository Reading		8, 15, 51	15, 16		
Expository Writing	19, 24		16, 18, 20, 25	20, 25	19, 21, 26
Letter Writing			22, 59	23, 59, 76	
Descriptive Reading	27, 28, 29, 30, 36	27, 30, 36, 63	30	27, 28	
Descriptive Writing	32, 33		31, 32, 37	31, 32, 37, 38	32, 38
Poetry	34, 35	34, 35	34, 35	34, 35	
Narrative Reading	29, 36, 42	29, 36, 39, 42, 43, 63	39	39	45
Narrative Writing	40		41, 44, 46, 49	44, 46, 47, 49	40, 45, 49
Fables		43	43, 46, 49	46, 47, 49	49
Persuasive Reading	52, 54	51, 54, 55, 58	51, 58		
Persuasive Writing	57	52	56, 59, 60, 61, 62	54, 55, 56, 59, 60, 61	57, 62
Essay Writing			25, 60, 61, 62	25, 60, 61, 77	26, 62
Research Report		63	64, 65, 67, 68,	67, 68, 69, 70, 71, 72, 74	65, 67, 68, 70, 73, 74
Using and Evaluating References/Sources	29		17, 52, 53, 64, 66	17	67
Revising				10, 14, 26, 32, 38, 49, 57, 62	
Proofreading				10, 14, 26, 32, 38, 49, 57, 62, 71	71
Formatting			73, 74	14, 22, 67, 73	
Bibliography			72, 74	72	72, 74
Publishing				26, 32, 38, 49, 50, 57, 62, 74	
Technology			53	48	48

Name ______________________ Date ______________

Words to the Wise: Complete Sentences

Do you remember the story "Goldilocks and the Three Bears"? Goldilocks found three bowls of porridge in the bears' cottage. One bowl was too hot, one bowl was not hot enough, and the other bowl was just right.

Groups of words are like that, too. A **sentence** is a group of words that forms a complete thought. It is just right. A **fragment** is a group of words that does not form a complete thought. It is not enough. A **run-on** is a group of words that runs two sentences together without using the words *and*, *or*, or *but*. A run-on is too much.

Good writers avoid writing fragments and run-ons. Good writers write complete sentences.

The three bears home. (fragment)
The three bears came home they found Goldilocks. (run-on)
The three bears came home. (sentence)

Read each group of words below. If the group is a sentence, write *sentence* on the line. If the group is a fragment or a run-on, rewrite the words to make a sentence.

1. Mother bears give birth to their cubs while hibernating.

   ____

2. Big bears that stand up to seven feet tall.

   ____

3. Grizzly bears mostly berries, nuts, and roots.

   ____

4. They are really brown bears whose hair gets "grizzled."

   ____

5. Grizzled means "silver-tipped" it happens as bears get older.

   ____

   ____

Write to Know Trivia

Mammals are the only animals with flaps around their ears.

0-7696-4026-5 *Writing*

Name ______________________________ Date ______________

Tools to Use: Punctuation

Three punctuation marks walked into a school. The period, the exclamation mark, and the question mark each had a reason to be there. Listen to each one.

Period: I belong at the end of a sentence. (makes a statement) **or**
Tell me what to do. (gives a command)

Question mark: Where do I belong? (asks a question)

Exclamation mark: Go right now! (gives a strong command) **or**
That is the best place to be! (shows strong emotion)

Read the following paragraph. Write the correct punctuation (**.**, **?**, or **!**) in each box.

Some Bear Facts

Alaska is home to many kinds of bears ☐ Can you name them ☐ Black bears, brown bears, and polar bears all live in Alaska ☐ It's no wonder Alaska is known as Bear Country ☐ Did you know that the brown bear is the largest land mammal in the world ☐ Well, it is ☐ The adult male can weigh up to 1,500 pounds ☐ That is amazing ☐ Don't get in its way ☐ A mother bear will fight to defend her cubs ☐ If you ever see a mother brown bear, you should do one thing ☐ Leave her alone ☐

Think About It!

Read each sentence in the paragraph again. Mark each one as follows:

If it makes a statement or gives a command, underline the sentence.

If it shows strong emotion or gives a strong command, draw a box around the sentence.

If it asks a question, circle the sentence.

Write to Know Trivia

Smokey the Bear was a real bear. He was rescued as a cub after a forest fire in New Mexico. He lived at the National Zoo in Washington, D.C., for many years.

0-7696-4026-5 *Writing*

Name ______________________ Date ______________

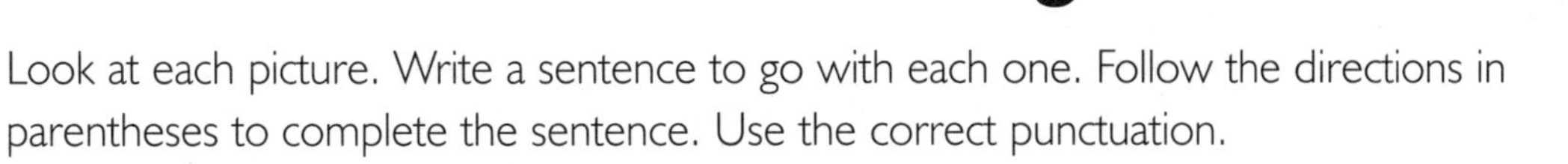

Your Turn! Writing Sentences

Look at each picture. Write a sentence to go with each one. Follow the directions in parentheses to complete the sentence. Use the correct punctuation.

1. (show a strong emotion) ______________________

2. (make a statement) ______________________

3. (give a strong command) ______________________

4. (ask a question) ______________________

5. (give a command) ______________________

Your Turn—Your Choice!

Write a complete sentence to go with each illustration.

6. ______________________

7. ______________________

8. ______________________

Write to Know Trivia

A male lion's mane gets darker as he gets older.

Name ______________________ Date ______________

Revise for the Prize

Read each poorly written group of words. Rewrite them to make clear, complete sentences. Be sure to use correct punctuation for each. Some sentences will have several possible answers.

1. Each baseball season, fans eat about 25 million hot dogs adults like mustard best.

   ____

2. Kids best like on their hot dogs ketchup.

   ____

3. Like foot-long hot dogs, too.

   ____

4. Many stories about the beginning of the hot dog.

   ____

5. Is really a sausage?

   ____

6. Hot dogs became baseball's favorite food in 1893 they were inexpensive and easy to eat.

   ____

7. Students Yale University bought from "dog wagons" their hot dogs.

   ____

8. The average person only about six bites to one hot dog.

   ____

Write It!

Describe your favorite hot dog. Use complete sentences.

____

____

Write to Know Trivia

The name "hot dog" comes from the original term "dachshund sausage," which means "little-dog sausage."

0-7696-4026-5 *Writing*

Name ______________________ Date ______________

Focus on Form: Paragraph Structure

Read the following paragraphs on skateboarding basics.

____

____

Look at the surface where you will be riding. Check it for holes, rocks, bumps, or cracks. The smoother the surface, the safer your ride will be. It is better to use a skate park than to skateboard in the street. Using ramps already built for skateboarders may cut down on accidents. Homemade ramps are often more dangerous.

____

____

____

____

Gear that is too tight limits movement and cuts down on blood and oxygen circulation. If your gear is too loose, it might slip out of place. A loose helmet, for example, might fall in front of your eyes and block your vision. A well-fitting helmet should not move when you shake your head. Besides a helmet, you should wear knee and elbow pads, gloves, and wrist guards.

____

____

Something is missing in both paragraphs above. Every paragraph has three parts. Those parts are a topic sentence, the body of the paragraph, and the concluding sentence.

The **topic sentence** tells what the paragraph is going to be about. The **body** includes sentences that support the topic sentence. If a sentence does not fit with the topic, it does not belong in the paragraph. Finally, the **concluding sentence** summarizes the paragraph.

Here are two topic sentences. Write each one at the beginning of its appropriate paragraph in the above examples.

Wearing correct skateboarding gear is important.
Skateboarding surfaces make a difference.

Now, reread each paragraph. Think of a sentence that summarizes the ideas. Write a concluding sentence on the last line of each paragraph.

Name ______________________ Date ______________

Tools to Use: Web Organizer

Jordan wanted to write about skateboarding moves. He read about them in magazines, books, and on Web sites. Jordan even interviewed some local skateboarders. He had a lot of information. How could he organize it? A web organizer is a good idea. Here is how it looks.

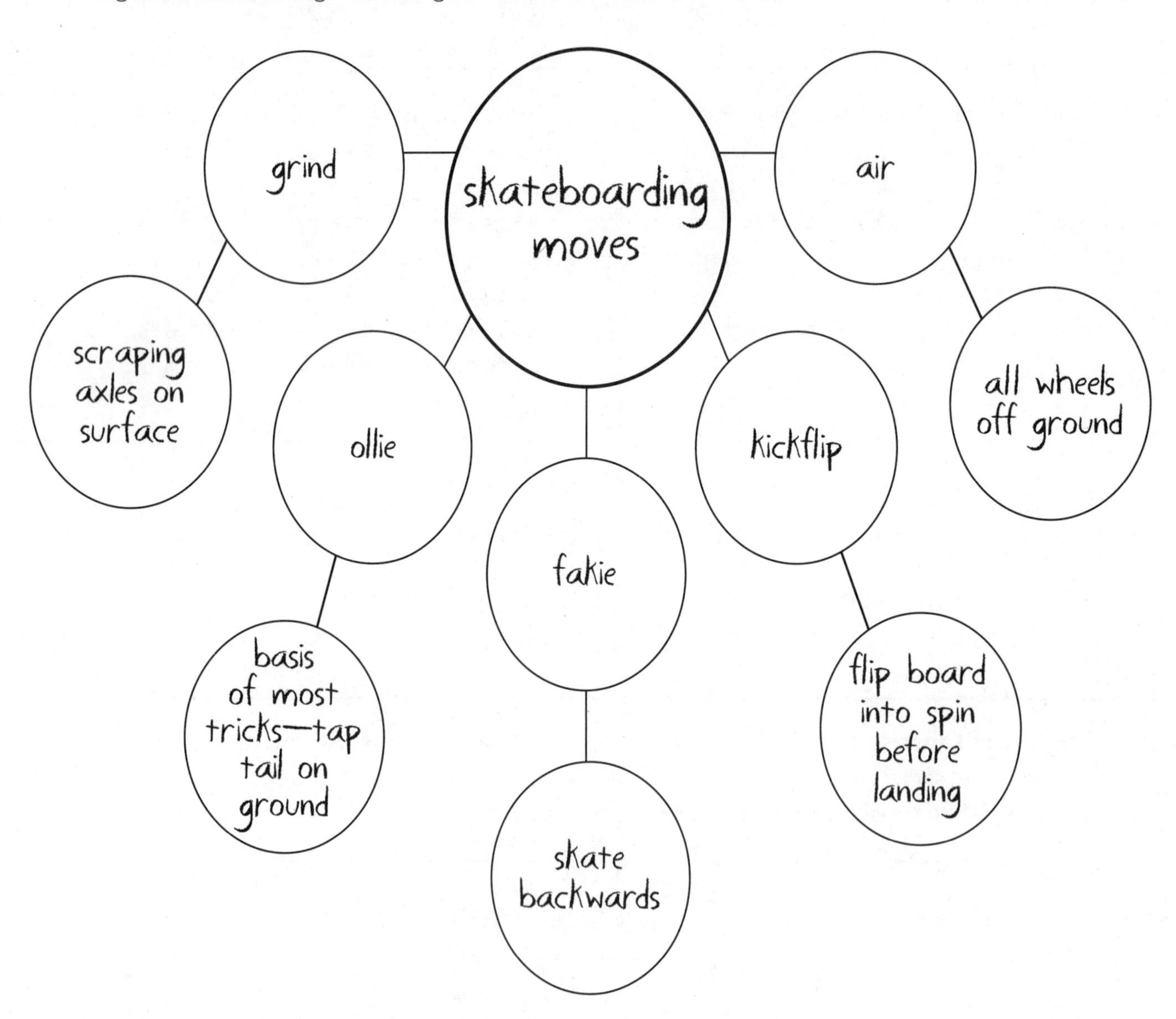

Do It Yourself!

Make your own web on another piece of paper. Write a main topic in the center. Create the rest of the web by adding notes and details about the topic. Use the example above to help you.

Write to Know Trivia

The "ollie" was named for the guy who first did it: Alan Ollie Gefland.

0-7696-4026-5 *Writing*

Name ______________________________ Date ______________

Your Turn! A Paragraph

Now it's time to turn the Skateboarding Moves web into a paragraph. Begin by writing a topic sentence. Use the topic in the center circle to do this. Write a sentence that tells what the paragraph will be about.

Topic sentence: ______________________________

Next, use the words in the connected circles to write sentences. These sentences must support the topic sentence.

Second sentence: ______________________________

____

Third sentence: ______________________________

____

Fourth sentence: ______________________________

____

Fifth sentence: ______________________________

____

Sixth sentence: ______________________________

____

Finally, think of a way to summarize the paragraph. Write a concluding sentence that brings it all together in a creative way.

Concluding sentence: ______________________________

Do It Yourself!

Use the web you created for your own topic to write another paragraph. Write the rough draft on a separate piece of paper. Be sure to include all three parts: the topic sentence, the body sentences, and the concluding sentence.

Write to Know Trivia

"Sidewalk surfing" started when surfers in California wanted to surf during bad weather. That's how skateboarding got its start!

Name ______________________ Date ______________

Revise for the Prize

Choose either the Skateboarding Moves paragraph or the one you wrote on your own. Now it is time to Revise for the Prize. Use this checklist when revising your work.

- ❑ All sentences are complete. There are no fragments or run-ons.
- ❑ All words are spelled correctly. I have checked the ones I didn't know.
- ❑ Each sentence has correct punctuation.
- ❑ The topic sentence is correct and clearly written.
- ❑ The topic sentence is supported by each sentence in the body.
- ❑ The concluding sentence summarizes the paragraph.
- ❑ My paragraph is interesting to read.

Make corrections to your rough draft. Rewrite your paragraph below.

____

____

____

____

____

____

____

____

____

____

Paragraph Perfection!

Write a final copy in ink or use the computer. Be sure to indent the first sentence of the paragraph. Double space the paragraph by skipping lines. Your computer program should be set on "double space." Use a file folder or small binder to hold your finished work. This will show your progress as you practice, practice, practice writing! Place the final draft of your paragraph in your portfolio.

Name ______________________ Date ______________

Write to Explain

The Real Deal

Read the following paragraph about plant allergies. Think about where you might find an article like this.

Which Itch Is This?

Some of the itchiest allergies come from three plants. They are poison ivy, poison oak, and poison sumac. Three out of four people are allergic to the oil on the plants' leaves, stems, and roots. When this oil gets on a person's skin, a rash appears within a few hours. The skin becomes very itchy and swollen. A mild rash usually itches for a week or more. A rash that lasts more than two or three weeks may need a doctor's attention. Unless you like to itch, stay away from these three plants!

Focus on Form

The purpose of this paragraph is to give information—to explain. It is an **explanatory** paragraph.

What is the paragraph explaining? ______________________

Read the paragraph again. Underline the topic sentence. Now list the facts that support the topic sentence.

____

____

____

Underline the concluding sentence twice.

Inquiring Mind?

What other information would you like to know about this itchy allergy? Write at least three questions you could use for research.

____

____

____

____

Write to Know Trivia

Burning a poison ivy plant does not keep you from getting a rash! The air can carry the oil particles, which are still enough to create an allergic reaction.

0-7696-4026-5 *Writing*

Name ______________________ Date ______________

Focus on Form: Topic Talk

Before you write an explanatory paragraph, you need a topic. The topic must be able to be explained clearly in a few sentences. If the topic is too broad, it needs more than one paragraph. Keep your topic focused.

Read each topic below. If the topic can be explained clearly in one paragraph, write **E** on the line. If it is too broad, write **B** on the line.

1. How to play soccer _____
2. What a soccer ball is made of _____
3. What a goalie does _____
4. The World Cup tournament _____
5. Setting up a soccer field _____
6. The history of soccer _____
7. How to steal the ball _____

Your Turn!

Think of a general subject you would like to write about. Write the subject on the line.

____

Now focus on some topics about the subject. Write them below.

____

____

____

Choose one of your topics that could be clearly explained in one paragraph. Circle it.

Write something you already know about the topic.

____

____

What other information will you need to find? ______________________________

Write to Know Trivia

During the 1960s and 1970s, Brazilian soccer hero Pelé played in 1,362 matches and scored 1,282 goals during his career.

Name ______________________ Date ______________

Tools to Use: Where to Find the Facts

Going to the right place is important. If you want to find a book, you go to a bookstore or library. If you want to eat a pizza, you go to a pizza restaurant. You can find what you are looking for, but only if you look in the right place.

Information is like that, too. You need to look in the right place. Information can be found almost everywhere! You can find facts in libraries, encyclopedias, dictionaries, atlases, special books, and magazines. These are called sources. You can find sources on the Internet or by interviewing experts. Where else can you find information?

____

____

Read each topic below. List sources you could use to find information on that topic. Write your answers on the lines. There is more than one correct answer for each one.

1. How to make an apple pie ______________
2. The best kinds of apples for cooking ______________
3. Facts about the biggest apple ever grown ______________
4. Who Johnny Appleseed was ______________
5. What scientists say about apples and health ______________
6. Where apples are grown ______________
7. What a crabapple is ______________
8. What kinds of insects can hurt apple trees ______________
9. How apple cider is made ______________
10. A fairy tale about a girl and a poisoned apple ______________

Go back to the topic you chose for your explanatory paragraph. List some sources you can use to find more information about that topic.

____

____

____

Write to Know Trivia

About eighty-four percent of a raw apple is water.

Name ______________________ Date ______________________

Tools to Use: Paragraph Ladder

A **paragraph ladder** is a good tool to use to organize your paragraph. Each part of the paragraph has a place on the ladder. The topic sentence goes at the top. The supporting details go in the middle. The concluding sentence goes at the bottom. The ladder on the left is an example.

Gather all of the information about your topic. Use the ladder on the right to organize your paragraph. You might not use all of the spaces or you may need to add some. Be sure that each detail supports the topic sentence.

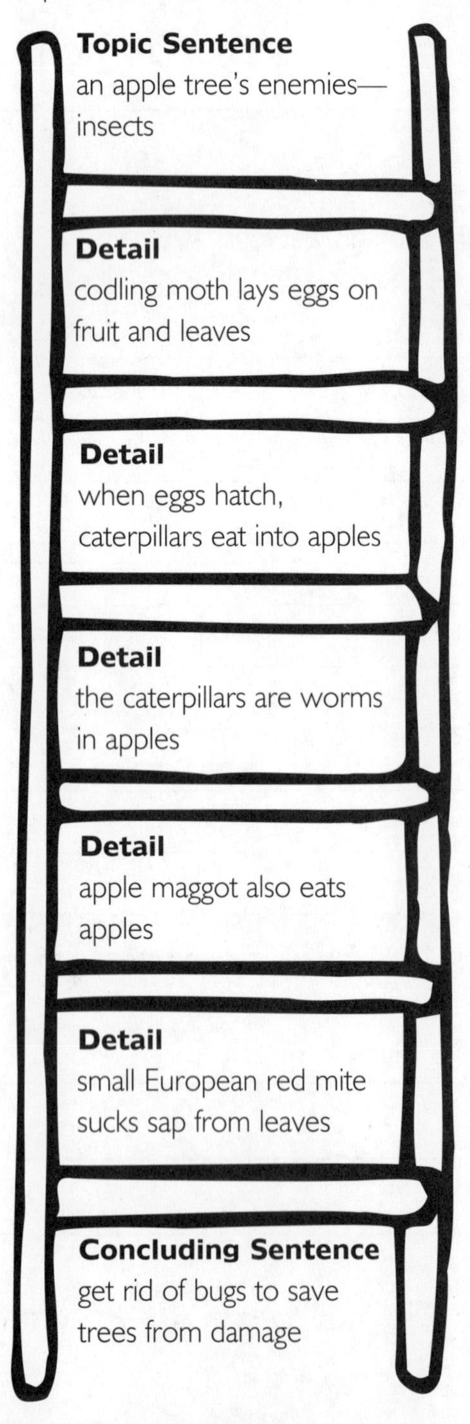

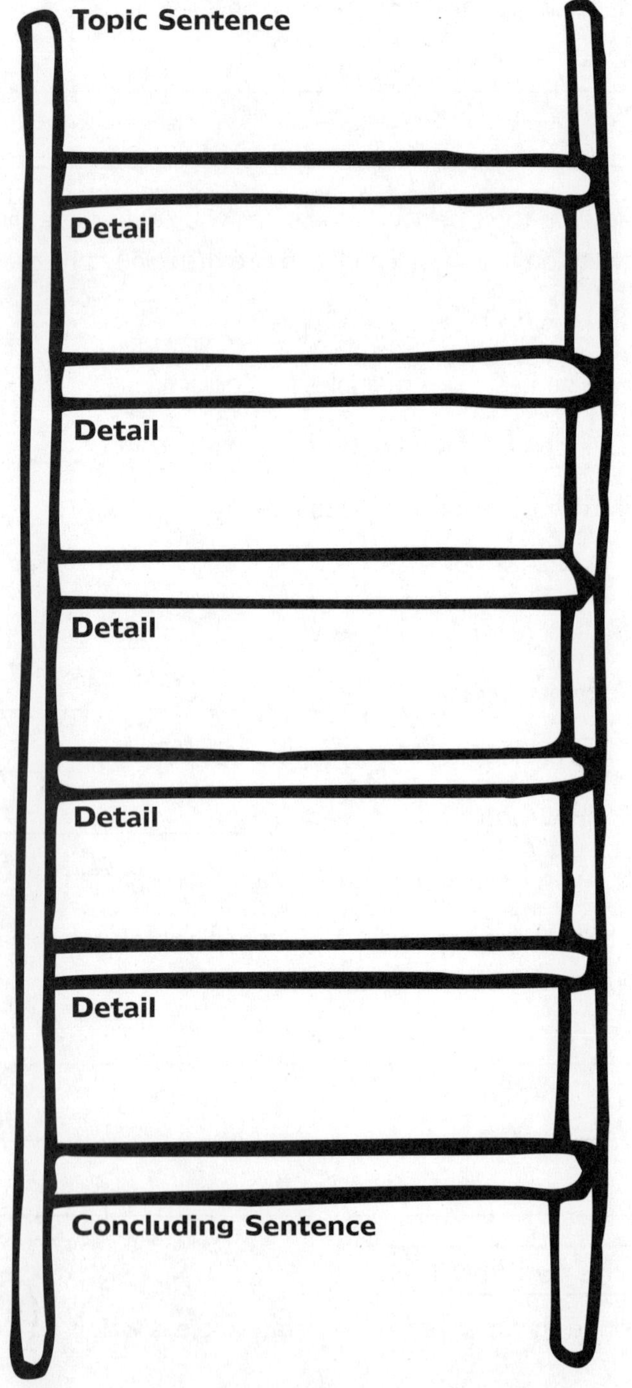

Name ______________________________ Date ____________________

Write to Explain

Words to the Wise: Transitions

Sentences sometimes need words that connect thoughts. **Transitions** help the reader. They are words or phrases that can tell the reader about:

- order—first, next, last
- position—nearby, next to, farther
- time—soon, suddenly, afterward

Transitional Words and Phrases

First	In addition	Soon	Therefore
Next	In the same way	Afterward	As a result
Finally	On the other hand	Before	For example
Then	Another	For that reason	Also
When	Instead	Nearby	
Last	Meanwhile	Next to	

Use the words in the box to fill in the blanks in the paragraph below. Make sure your choice makes sense as you read the paragraph.

Apple trees have some insect enemies. ____________________ the codling moth likes to lay eggs on the fruit and leaves. ____________________ the eggs hatch, the new caterpillars eat their way into the apples. The worm in an apple is really this caterpillar! ____________________ the apple maggot eats the apples. ____________________ small insect, the European red mite, sucks the sap from the tree's leaves. They cause a lot of damage. ____________________ it is important to get rid of these pests.

Name ______________________ Date ______________

Your Turn! Expository Paragraph

It's time to use the paragraph ladder you created for your explanatory paragraph! Look at the information you wrote. Make changes if you need to. Remember, you are *explaining* something. Ask yourself:

- ✎ Is the topic sentence clear?
- ✎ Do the details support the topic sentence?
- ✎ Do I have enough information?
- ✎ Do I have the details in some kind of order?
- ✎ Does the concluding sentence pull the paragraph together?

If you answered "no" to any of the questions, redo your ladder. Once your ladder is ready, use the parts to write your paragraph. Indent the first sentence. Make sure each sentence is complete—no fragments or run-ons. Use transitional words and phrases as needed.

0-7696-4026-5 *Writing*

Name ______________________________ Date ______________

Write to Explain

Tools to Use: Commas

Punctuation is a tool. Each mark does something special. The period ends a sentence or shortens words. The question mark turns a sentence into a question. An exclamation mark adds excitement to a sentence.

The **comma** has a purpose, too. A comma creates a short pause between two words. Look at the "comma-less" sentence below. Does it make sense? Are you sure about what the sentence means?

Teresa the teacher wants to give us paper nametags and markers.

Let's add some commas and see how it changes the meaning.

1. Teresa, the teacher, wants to give us paper nametags and markers.
2. Teresa, the teacher, wants to give us paper, nametags, and markers.
3. Teresa, the teacher wants to give us paper, nametags, and markers.

How is each one different?

____

____

____

Correctly add commas to the sentences below.

1. When I was little I had a fish a hamster and a rabbit.
2. I won the first race but then I lost the second.
3. Yes it was a green black and silver insect.
4. Will you bring your brother Mary?
5. Mr. Andrews the mayor came to dinner.

Revise for the Prize!

Look over the explanatory paragraph you have been working on. Did you use commas correctly? Revise the paragraph and fix any problems, such as spelling, punctuation, capitalization, fragments, and run-ons.

Write your final copy in ink or use a computer. Place the final copy in your Personal Portfolio. Congratulations! You wrote an expository paragraph!

Write to Know Trivia

During the Middle Ages, commas were written as a slash mark (/).

Name ______________________ Date ______________

Write to Explain

Focus on Form: Letter Format

Stacy found this letter in a bottle. It was torn to pieces. Put the letter back together so that each part is where it belongs. Write it out on the letter form below.

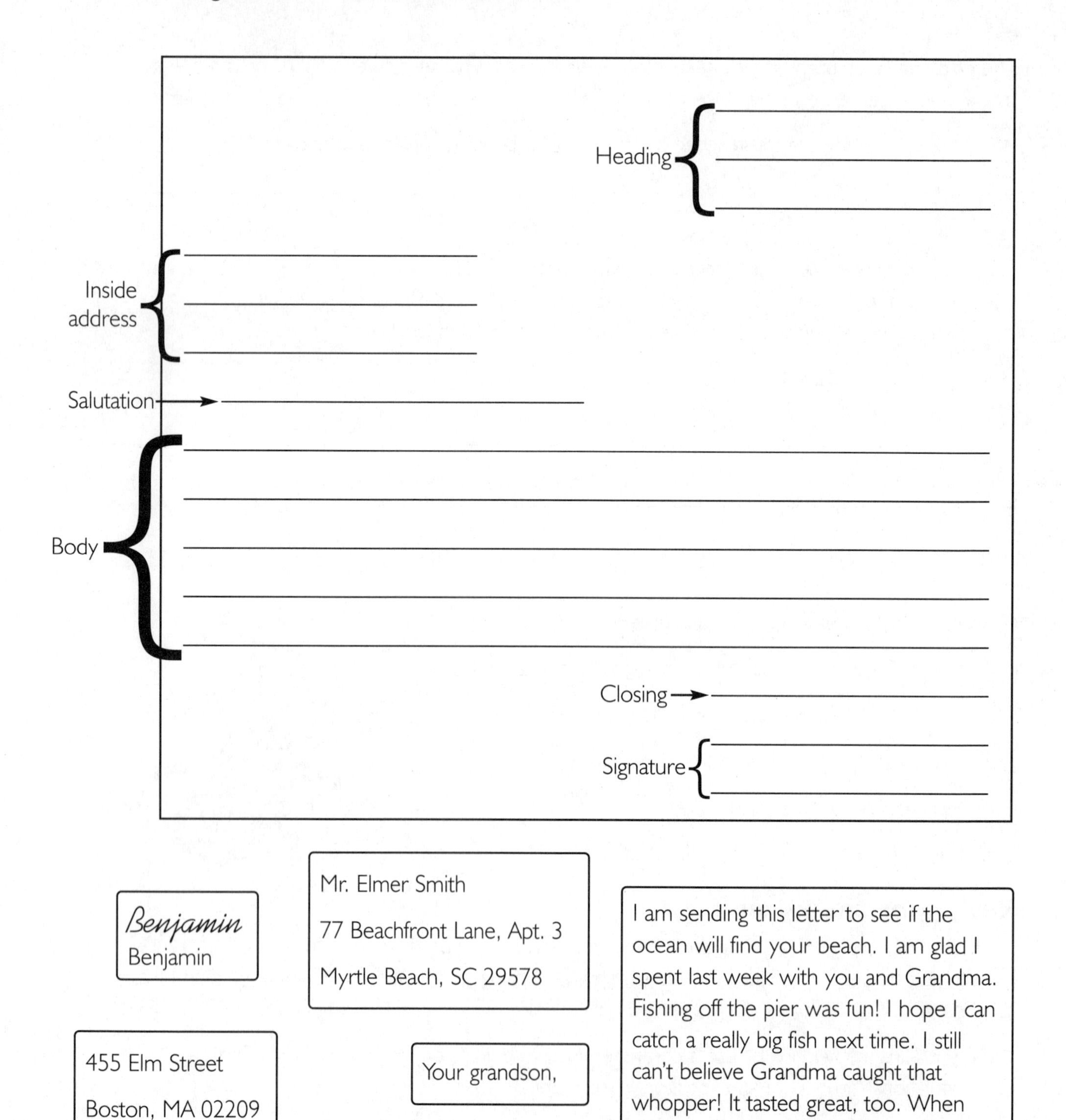

Benjamin
Benjamin

Mr. Elmer Smith
77 Beachfront Lane, Apt. 3
Myrtle Beach, SC 29578

I am sending this letter to see if the ocean will find your beach. I am glad I spent last week with you and Grandma. Fishing off the pier was fun! I hope I can catch a really big fish next time. I still can't believe Grandma caught that whopper! It tasted great, too. When you go fishing again, think of me.

455 Elm Street
Boston, MA 02209
June 22, 2004

Your grandson,

Dear Grandpa,

Name ______________________________ Date ______________

Your Turn! Expository Letter

Sometimes you need to write a letter that explains something to someone. For example, you might want to explain how to get to a place. Or, you might want to explain why you like something. You might even explain a problem and ask for a solution.

Write a letter to someone you know. Explain something to him or her. Be sure to follow the correct format for a letter. It should have:

- heading (your address and the date)
- inside address (name and address of the person who will receive the letter)
- salutation (Dear _______ and follow this with a comma)
- body (the paragraph(s) of the letter)
- closing (Yours truly or Sincerely and follow this with a comma)
- signature (your signed name above your printed or typed name)

Write a rough draft of the letter's body paragraph(s) here.

Revise for the Prize!

Look over your letter. Work with a partner to check for errors. Fix any problems you find—spelling, punctuation, capitalization, fragments, and run-ons.

Write your final copy in correct letter form. Place the final copy in your Personal Portfolio or send the letter to the person in your salutation.

Name ______________________ Date ______________________

Words to the Wise: Idioms

Our language contains some interesting sayings. One type of saying is called an **idiom**. An idiom is an expression used by a common group of people. It is unique either by grammar rules or by meaning.

Example: Cho is "hitting the books" for her test tomorrow.

This does not mean Cho is striking her books with her hands. The phrase is commonly used to describe someone studying hard to prepare for a test.

Read each of the idioms below. Write what you think each one means.

1. Birds of a feather flock together. ______________________

____

2. The early bird gets the worm. ______________________

____

3. Don't count your chickens before they hatch. ______________________

____

4. eat like a bird ______________________

____

5. ugly duckling ______________________

____

Your Turn!

How many idioms do you know? Work with a partner to create a list of idioms. Share your list with classmates. Discuss how these phrases or sayings can mean different things to different groups of people.

Write to Know Trivia

A group of geese on the ground is called a *gaggle*. A group of geese in the air is called a *skein*.

0-7696-4026-5 *Writing*

Name ______________________________ Date ______________

Your Turn! Expository Essay

An **essay** is made up of at least three paragraphs: an introductory paragraph, the body paragraph or paragraphs (which support the main idea of the first paragraph), and a concluding paragraph. Below is an example of how to build an essay.

Start with a topic.

Superman, the superhero

Create a main idea, or **focused topic**, for your first paragraph.

Superman has many superhuman skills.

He can fly.
He is stronger than all other people.
He has super senses.

Write some details that support this statement. Add examples for each. These are your second, third, and fourth body paragraphs.

He can fly.

Gets from place to place quickly
Can go into space
Can go wherever he wants

He is stronger than all other humans.

Can lift things
Cannot be hurt by others
Can protect people

He has super vision and hearing.

Has x-ray vision
Can see through buildings
Can hear through walls
Can hear over distances

Summarize the information for your concluding paragraph.

Super-human skills make Superman a hero.

Uses super skills to help people in need
Sees danger and responds quickly

Use the outline above to write a five-paragraph essay. Be sure to have a clear topic sentence for each paragraph. Write complete sentences. Write a rough draft on a separate piece of paper.

Name ______________________ Date ______________________

Revise for the Prize

Revise your rough draft. Use this checklist:

- Did I follow my outline?
- Does each paragraph have a clear topic sentence?
- Is every sentence complete (no fragments or run-ons)?
- Are all words spelled correctly?
- Did I capitalize correctly?
- Did I use commas and other punctuation correctly?
- Did I indent the first sentence of each paragraph?
- Is my essay neatly written or printed out?

Write your final draft in ink on your own paper or use a computer to print it out. Then place it in your Personal Portfolio. Congratulations! You have written a "super" essay!

Here's What I Think...

Use the lines below to write a paragraph. Explain how easy or difficult it was to write the essay. Include a topic sentence. Give examples. Write a concluding sentence.

____

____

____

____

____

____

____

____

____

____

____

Write to Know Trivia

In the movie *Spiderman*, about 150 spiders were used. The movie's bug consultant, Steve Kutcher, stated that no spiders were harmed.

Name ______________________________ Date ______________

Write to Describe

The Real Deal

Read the following description. Think about where you might find a paragraph like this.

A tall, furry-horned fellow blinks his long eyelashes as he eats. His long, agile tongue wraps around leaves and pulls them into his mouth. He chews, swallows, and burps the food back up. Then he chews and swallows it again. As he eats, he is hidden among the tall trees. His large spots help him blend in. Although his neck is very long, it has only seven vertebrae, just like other mammals. He stands about 18 feet tall and weighs as much as 4,000 pounds. He gallops on strong hooves, reaching speeds of more than 35 miles per hour. This African animal is the tallest of all land mammals. It is the tall, tall giraffe!

This is a **descriptive** paragraph. Its purpose is to give the reader a picture of something—details about how it looks, feels, sounds, smells, or tastes. Read the paragraph again and answer the following questions.

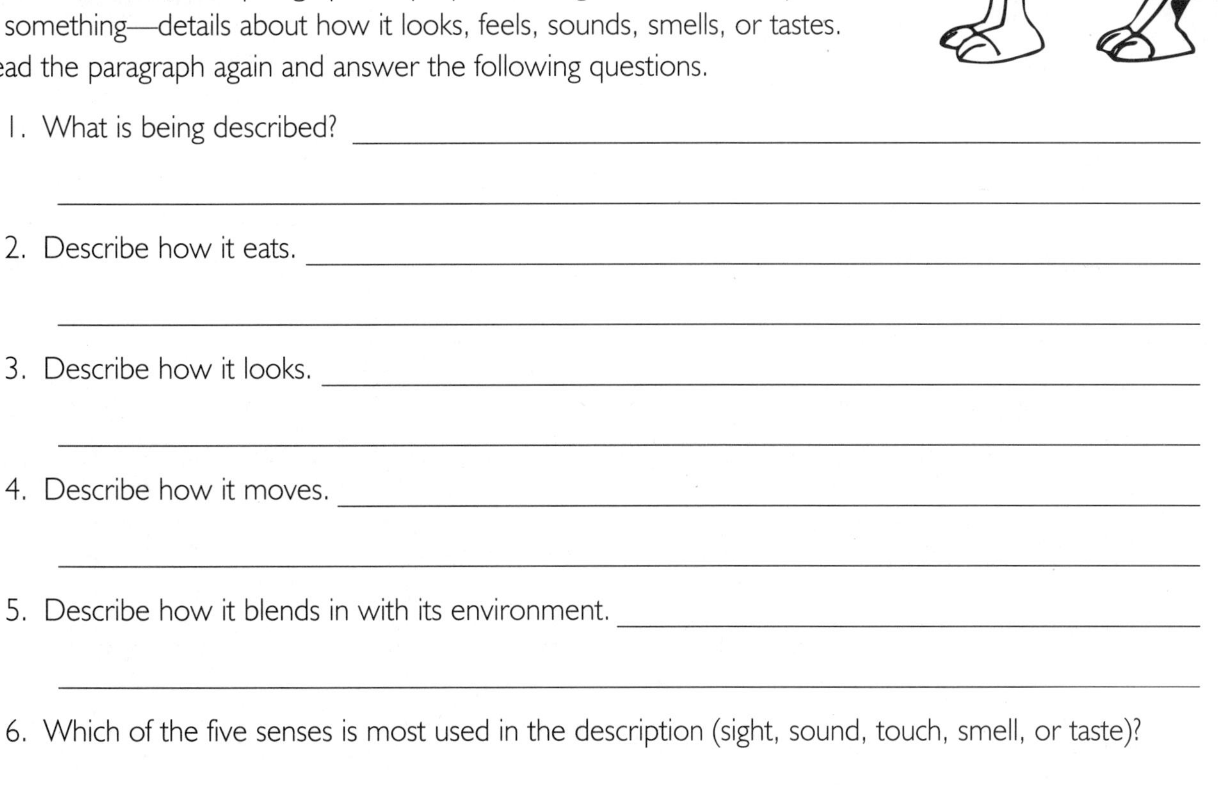

1. What is being described? ______________________________

____

2. Describe how it eats. ______________________________

____

3. Describe how it looks. ______________________________

____

4. Describe how it moves. ______________________________

____

5. Describe how it blends in with its environment. ______________________________

____

6. Which of the five senses is most used in the description (sight, sound, touch, smell, or taste)?

____

Write to Know Trivia

A giraffe can clean its own ears with its twenty-inch-long tongue.

Name ______________________________ Date ______________

Words to the Wise: Strong Words

Not all words are created equal. Some words are general. Some are **specific**. The more specific a word is, the stronger it becomes. The best words in writing are those that give your reader a clear picture of your subject. Read each sentence below. Look at the underlined word. Is it a weak word or a strong word? Circle your choice.

1. The girl <u>spoke</u> to her friend.	weak	strong
2. The teenager <u>screamed</u> at her brother.	weak	strong
3. The <u>mansion</u> cast a huge shadow on the town below.	weak	strong
4. The house on the hill was <u>big</u>.	weak	strong
5. I <u>cried</u> all night long.	weak	strong
6. I <u>sobbed</u> after he told me his story.	weak	strong
7. Bob <u>made</u> the basket easily.	weak	strong
8. Bob <u>slammed</u> the basketball through the hoop.	weak	strong

Strong words make your writing more interesting. You can use strong nouns, verbs, and modifiers to improve your writing. Read each word below. Write a stronger word or a specific example that could be used instead.

9. bird ______________________________
10. car ______________________________
11. walked ______________________________
12. small ______________________________
13. dark ______________________________
14. drove ______________________________
15. said ______________________________
16. Read the following sentence. Circle each strong word that gives you a clear image.

A tall, furry-horned fellow blinks his long eyelashes as he eats.

Write to Know Trivia

The giant squid has the biggest eyes of any animal. Each eye is about the size of a volleyball!

Name ______________________________ Date ______________

Write to Describe

Tools to Use: Thesaurus

Words, words, words! What can you do when you can't find the word you want? Check out the **thesaurus**. It is arranged like a dictionary. (You might have to look up the root word if it is a verb. For example, thought = think, pulled = pull, looking = look)

Practice using a thesaurus to find a stronger word for each of the underlined words in the poem below. Write the word on the line provided.

I <u>thought</u> and thought ____________
And thought some more
Erased my words
'Til my paper <u>tore.</u> ____________
I thought so hard
My brain got sore.
I <u>threw</u> my pencil ____________
Out the door.
What was that word
I was <u>looking</u> for? ____________
My <u>teacher</u> said, ____________
"I'll help you, Doris.
What you <u>need</u> ____________
Is a thesaurus!"

"What is that?
Like a dinosaur?
An <u>ancient</u> place? ____________
A <u>piece</u> of lore?" ____________
My teacher <u>smiled</u>, ____________
And from a drawer,
She <u>pulled</u> a book ____________
I'd seen before.
A perfect <u>book</u> ____________
With words <u>galore!</u> ____________
She left it out,
Said it was for us,
This <u>wonderful</u> book, ____________
The great thesaurus!

Write to Know Trivia

The word "encyclopedia" comes from two Greek words that mean "a circle of learning."

0-7696-4026-5 *Writing*

Name ______________________ Date ______________

Tools to Use: Sensory Details

Using your five senses can make your writing realistic and vivid. The more you can describe a person, place, or thing using your senses, the better. These kinds of descriptions are called **sensory details**.

Read the paragraph below. Look for words or phrases that give sensory details. Underline them.

It was the biggest plant I'd ever seen. Standing more than six feet tall, it pointed straight up. The plant looked like a giant tan spear wrapped in lettuce leaves. The "leaves" were the flower getting ready to open. As soon as it opened, the people around it all moaned in disgust. The plant earned its nickname, Mr. Stinky. The five-foot bloom produced the worst smell you could imagine. It smelled like a rotting elephant. Another name for a plant like Mr. Stinky is "corpse plant." Now I know why!

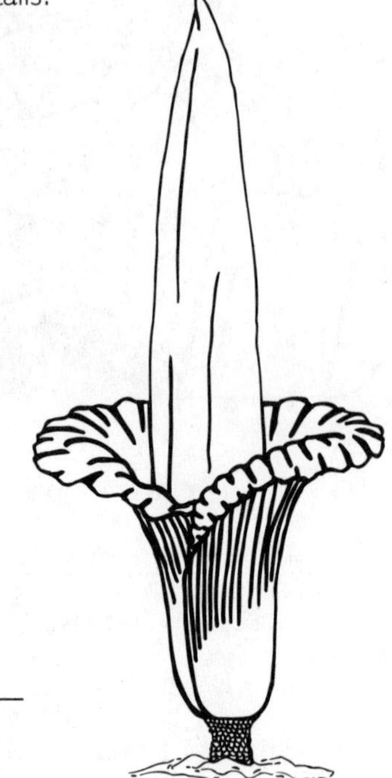

Which of the five senses are used in the paragraph?

____

Think of something you would like to describe. It can be a place or a thing. It could even be a person or an activity. Write the name of it here.

____

When you think of it:

What do you see? ______________________________

What do you hear? ______________________________

What do you feel? ______________________________

What do you smell? ______________________________

What do you taste? ______________________________

Write to Know Trivia

The Guinness Book of World Records (2004) named the corpse plant (*Amorphophallus titanium*) the smelliest flower on Earth. You can smell it more than half a mile away!

0-7696-4026-5 *Writing*

Name ______________________________ Date ______________

Your Turn! Descriptive Paragraph

Do you remember what makes a good paragraph? Unscramble the words to fill in this checklist.

1. Make sure it has a ______________ sentence. cpito
2. Write sentences that ______________ the first sentence. protsup
3. Use ______________ details to make your writing descriptive. osenyrs
4. End the paragraph with a ______________ sentence. iconlcdugn
5. Make sure each sentence is ______________, not a fragment or a run-on. epcometl
6. Be sure to ______________ the first sentence of the paragraph. tdenin

Use sensory details to write your own descriptive paragraph. Write a rough draft of your paragraph on the lines below.

____

____

____

____

____

____

____

____

____

____

____

____

____

____

0-7696-4026-5 *Writing*

Name ______________________________ Date ______________

Revise for the Prize

Review your descriptive paragraph. How many of the five senses are being used to describe your subject? Check each one that is used.

☐ sight ☐ sound ☐ touch ☐ smell ☐ taste

Now look at your words. Are they strong? Can you choose better words to describe your subject? List the words from your paragraph that you think you can improve.

____

____

____

____

List some new words you could use to replace the words above. You may use a thesaurus to help you.

______________________ ______________________

______________________ ______________________

______________________ ______________________

______________________ ______________________

Revise your paragraph. Write a final draft in ink or use a computer and printer. Put the final draft in your Personal Portfolio. Congratulations! You have written a descriptive paragraph!

____

____

____

____

____

Write to Know Trivia

A snail has two pairs of tentacles on its head. The longer pair has the eyes. The shorter pair is used for smelling and feeling where it's going.

0-7696-4026-5 *Writing*

Name ____________________ Date ____________

Words to the Wise: Connotation

A word can have two kinds of meanings. It has a dictionary definition meaning. It also has a meaning that affects how you feel or think about the word. Here is an example of three words that mean basically the same thing in the dictionary.

large	great	fat

What is the difference between these three words? Use each word in a sentence to show the difference in meanings.

1. large ____________________
2. great ____________________
3. fat ____________________

A word's **connotation** is the emotion or idea we feel when we think of that word. It can be positive or negative. Some words have a neutral connotation; we don't feel one way or the other about that word.

Look at each word below. Write on the line if its connotation is positive, negative, or neutral.

4. thin ____________________
5. slender ____________________
6. skinny ____________________
7. skeleton-like ____________________
8. Which of those four words has the strongest connotation to you? ____________________

Circle the word that has the strongest connotation in each group below. Use a dictionary to look up those words you do not know.

9. crude	unmannerly	rude
10. powerful	mighty	strong
11. fast	quick	hasty
12. wealth	riches	money

Write to Know Trivia

The heart of a blue whale is the size of a small car. It weighs about two thousand pounds!

Name ______________________ Date ______________

Focus on Form: Haiku

The Japanese people have written poetry for a long time. They have used a simple form, called **haiku**, to describe things in nature.

Haiku uses few words, but it is very descriptive. It has only three lines. The first line has five syllables. The second line has seven syllables, and the last line has five again. Here is an example.

Rushing water flows
Over, down a sudden edge
Crashing waterfall

Circle the descriptive words in the haiku above.

Which words help you see the image? ______________________

Which words help you hear? ______________________

Which words help you feel? ______________________

Now write your own haiku poems. Use descriptive words. Count the syllables carefully!

______________________ (5)
______________________________ (7)
______________________ (5)

______________________ (5)
______________________________ (7)
______________________ (5)

Review your haiku poems. Did you use sensory words to create your images? Are there stronger words you can choose? Look in the thesaurus for any words you feel could be more descriptive. Write your final versions and add them to your Personal Portfolio.

0-7696-4026-5 *Writing*

Name ______________________________ Date ______________

Focus on Form: Free Verse

Free verse is another form of poetry. It is "free" because there are no rules about rhyming, rhythm, or syllables. You can write what you feel, what you see, or what you imagine. You can even make your own rules about punctuation and capitalization!

Here is a famous free verse poem that describes a simple scene.

so much depends
by William Carlos Williams
(1883–1963)

so much depends
upon

a red wheel
barrow

glazed with rain
water

beside the white
chickens

Which of the five senses is most used in this poem?

____

Describe the scene in your own words.

____

____

____

What is a wheelbarrow used for? What do you think the poet means when he writes, "so much depends upon" a red wheelbarrow?

____

____

____

____

____

Think of a scene you can describe. Write your own free verse poem to describe it. Write it on another piece of paper and add it to your Personal Portfolio.

Write to Know Trivia

In still air, raindrops fall between 7 and 18 miles per hour. The speed depends on the size of the raindrop.

0-7696-4026-5 *Writing*

Name ______________________ Date ______________

Read and Write: Literature Response

This paragraph from a longer true story uses description. It is from *My First Buffalo Hunt* by Chief Luther Standing Bear.

> Away I went, my little pony putting all he had into the race. It was not long before I lost sight of my father, but I kept going just the same. I threw my blanket back, and the chill of the autumn morning struck my body, but I did not mind. On I went. It was wonderful to race over the ground with all these horsemen about me. There was no shouting, no noise of any kind except the pounding of horses' feet. The herd was now running and had raised a cloud of dust. I felt no fear until we had entered this cloud of dust and I could see nothing about me—I could only hear the sound of feet. Where was Father? Where was I going? On I rode through the cloud, for I knew I must keep going.

Underline descriptive words or phrases in the paragraph that stood out to you. Write about the author's use of sensory details. Was it effective? Could you feel as if you were there with him?

____

____

____

____

____

____

____

____

____

____

Write to Know Trivia

When Native Americans hunted buffalo, or bison, they used every part of the animal. They even used the rough side of the animal's tongue as a hairbrush!

Name ______________________________ Date ______________

Your Turn! Describe a Friend

Choose a friend to describe—a human friend or an animal friend. Begin by making a list of details about what he or she looks like, acts like, sounds like, and other traits that describe your friend.

Looks	Acts	Sounds	Other Traits

Now write a descriptive paragraph about your friend. Follow the correct paragraph form, using topic and concluding sentences.

Write to Know Trivia

About seventy percent of pet owners sign their pets' names on greeting cards, according to the American Animal Hospital Association.

Name ______________________________ Date ______________

Revise for the Prize

Share your Describe a Friend rough draft with someone who knows your friend. Read the paragraph to him or her. Do not tell whom the paragraph is describing. See if this person can tell who your friend is. If he or she cannot tell, then revise your paragraph to better describe your friend. If your friend can tell, then check your paragraph using the final draft checklist.

Write your final draft below. Illustrate or decorate a frame around your paragraph.

____

____

____

____

____

____

____

____

____

____

____

____

____

Write to Know Trivia

In 1935, the United States Congress proclaimed the first Sunday in August as National Friendship Day. It has been celebrated on that day every year since.

0-7696-4026-5 *Writing*

Name ______________________________ Date ______________

The Real Deal

Read the following story. Think about where you might find writing like this.

I heard the sirens scream through the darkness. It was after midnight, so I rolled over to go back to sleep. The sound of the sirens grew louder and louder. Suddenly, I saw red lights flashing through my open window. The fire department was at my house!

I ran down the stairs to see what was wrong. My heart pounded as I thought of my house burning down. I sniffed, thinking I'd smell smoke for sure. There was no smoke. Then I heard laughter outside. The sirens had been turned off.

As I looked out the back door, I saw a fire hose stretched across my backyard. I followed the hose into the darkness. My dad and mom stood behind our garage, laughing with the two firemen. My dad was holding the hose, pointing it toward a big hole. The water rushed out and disappeared into the opening.

"Surprise!" shouted my parents.

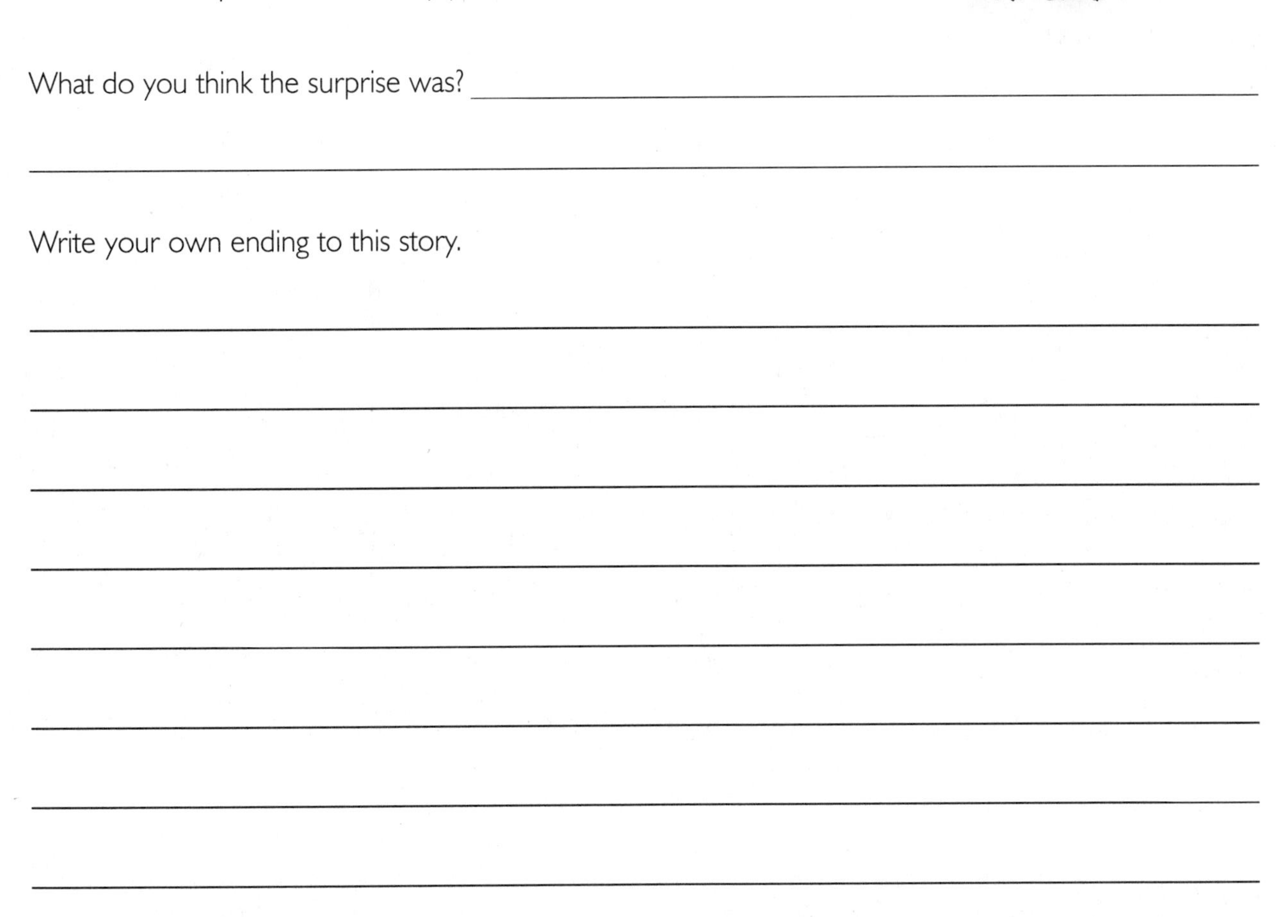

What do you think the surprise was? ______________________________

____

Write your own ending to this story.

____

____

____

____

____

____

____

____

____

0-7696-4026-5 *Writing*

Name ____ Date __________________________

Words to the Wise: Point of View

Every story has a storyteller, or **narrator**. In the Real Deal story on page 39, the narrator is the main character. This is called **first-person point of view**. The narrator is the "I" in the story. First-person pronouns are *I, me, my, mine, we, us, our,* and *ours*.

When the narrator is not in the story, it is written in **third-person point of view**. The characters may be named and are referred to as *he, she, they,* or *it*. Third-person pronouns are *he, him, his, she, her, hers, it, its, they, them, their,* and *theirs*.

When you write a story, decide which point of view you want to use. Then use the same point of view through the whole story.

Change the Real Deal story to third-person point of view. Choose a name for the main character. Change the first-person pronouns to third-person pronouns.

____________ heard the sirens scream through the darkness. It was after midnight, so ____________ rolled over to go back to sleep. The sound of the sirens grew louder and louder. Suddenly ____________ saw red lights flashing through ____________ open window. The fire department was at ____________ house!

____________ ran down the stairs to see what was wrong. ____________ heart pounded as ____________ thought of ____________ house burning down. ____________ sniffed, thinking ____________'d smell smoke for sure. There was no smoke. Then ____________ heard laughter outside. The sirens had been turned off.

As ____________ looked out the back door, ____________ saw a fire hose stretched across ____________ backyard. ____________ followed the hose into the darkness. ____________ dad and mom stood behind ____________ garage laughing with the two firemen. ____________ dad was holding the hose, pointing it toward a big hole. The water rushed out and disappeared into the opening.

"Surprise!" shouted ____________ parents.

0-7696-4026-5 *Writing*

Name ______________________________ Date ______________

Tools to Use: Story Planning

Every story, or **narrative**, has parts. It has a beginning, middle, and end.

It has characters, a setting, and a plot. The characters are the people or things that are doing the action of the story. The setting is the place and time the story occurs. The plot is the action of the story.

The plot shows a problem, events that build around the problem, and a solution. The problem should be solved near the end of the story.

Use the planner below to organize a story you would like to write. For each of the story elements, consider the bullet points to help develop your details.

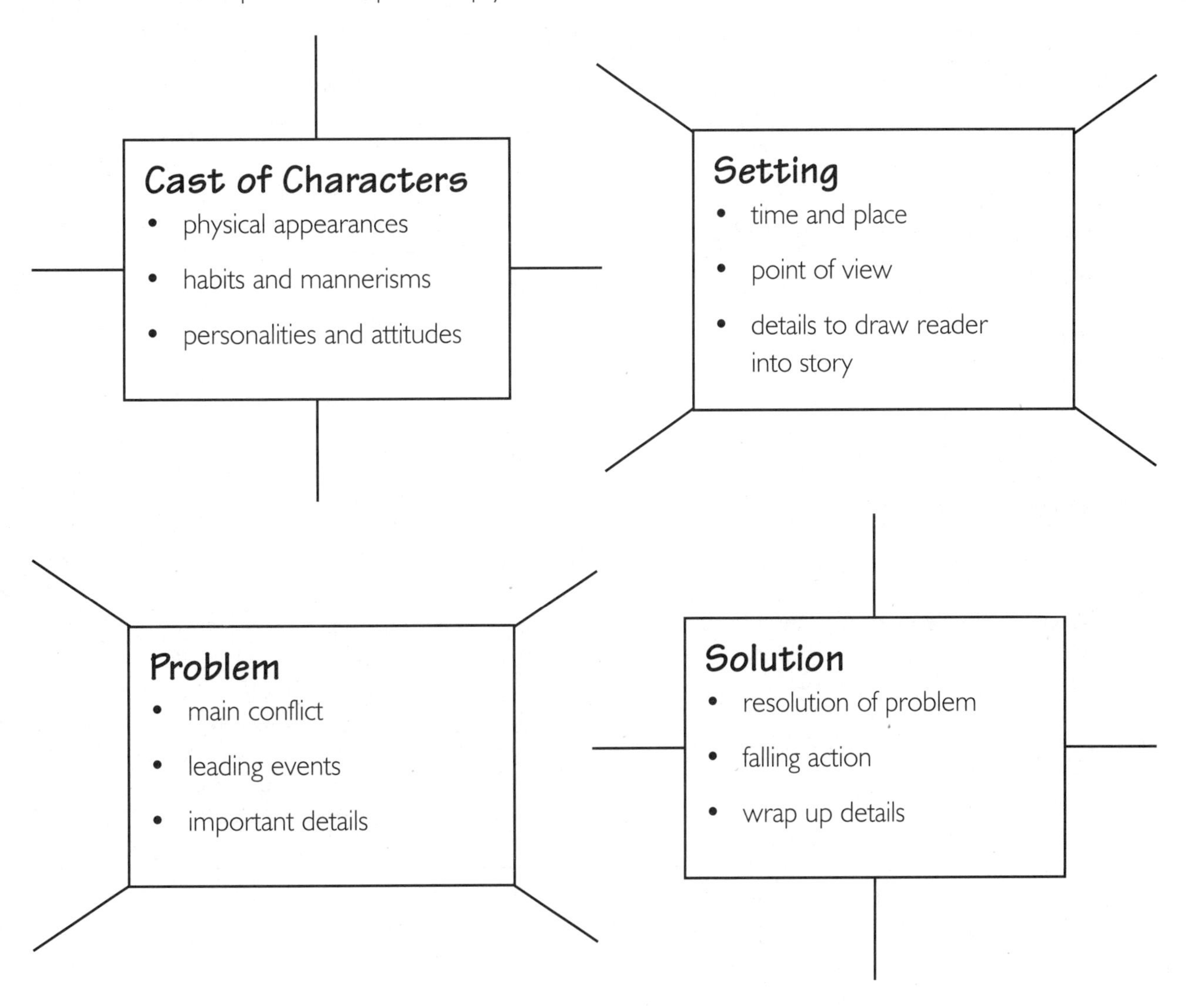

Use another piece of paper to write your rough draft. Work with a partner to revise and strengthen your story. Publish it and add it to your Personal Portfolio when finished.

Name ______________________________ Date ______________

Write to Narrate

Read and Write: Literature Response

Read the following true story. Then answer the questions about it.

What would you do if you found $20,000? Sixteen-year-old Dylan and his father had to decide for real. On July 4, 2004, Dylan and his dad were driving to their farm in Milford, Iowa. A vehicle drove past them, and suddenly paper was everywhere! The two men thought it was confetti. Then they noticed it was money! They stopped to pick it all up. One of the pieces of paper had a phone number on it. Dylan and his dad didn't even think twice. They knew the money belonged to someone else.

They called the number and got the address of the owner. Dylan walked up to a woman who was sitting outside her home. Jody Gardner was so surprised when Dylan gave her the money! She didn't realize that she had left her wallet on the roof of her van. The money had been her inheritance after her father's recent death. She offered them a $100 reward. They went home with only a smile instead.

1. What title would you give this story? ______________________________
2. Is this story written in first-person or third-person point of view? ______________
3. Who are the main characters? ______________________________

____

4. What is the place setting? ______________ What is the time setting? ______________
5. What is the main problem the characters must solve? ______________________

____

6. What events happen? (Name them in the order in which they happen.)

____

____

____

____

____

____

7. How is the problem solved? ______________________________

____

8. What does the ending mean? ______________________________

Name ________________________________ Date ________________

Focus on Form: Narrative Paragraph

A **narrative paragraph** can tell a made-up story, too. It has a beginning, middle, and end, like any story. Here is an example, adapted from a fable by Aesop, called "The Ant and the Dove."

An Ant went to the bank of a river to quench its thirst. It fell in and was carried away by the rush of the stream. A Dove, sitting on a tree near the water, plucked a leaf and let it fall into the stream close to the Ant. The Ant climbed onto it and floated safely to the bank. The same day a man came and stood under the tree. He was a bird-catcher, and he held out a twig for the Dove, which sat in the branches. The Ant, seeing what the man was up to, bit him on the foot. In pain, the bird-catcher threw down the twig. The noise made the Dove safely fly away.

1. Who are the characters in the narrative? ________________________________

____

2. What is the setting? ________________________________

____

3. There are two problems in the story. What are they? ________________________________

____

____

____

4. How are the problems solved? ________________________________

____

____

____

5. A fable ends with a moral, or lesson learned by the story. Aesop wrote that the moral of this story is "One good turn deserves another." What does that mean?

____

____

____

Name ______________________ Date ______________

Your Turn! Narrative Paragraph

Choose one of the following topics. Write your own narrative paragraph. The story can be true or made-up. Choose your point of view carefully. Be sure your paragraph has a beginning (topic sentence), middle (supporting sentences), and end (concluding sentence).

an embarrassing moment	an unforgettable day
the luckiest time	the worst birthday ever
a funny thing that happened	the scariest night ever

0-7696-4026-5 *Writing*

Name ____ Date __________________________

Words to the Wise: Talk about Punctuation

Stories have characters. Characters sometimes talk, and this is known as **dialogue**. In writing, dialogue uses punctuation in specific ways. Notice the **quotation marks** (“ ”), commas (,), and ending punctuation in these samples. In dialogue, ending punctuation belongs inside the last quotation mark. It's important to begin a new paragraph each time another person speaks, so the reader won't get confused.

> **“**Be very**,** very quiet**,”** whispered José**.**
>
> **“**Why should we**?”** asked Angela**.**
>
> José answered**, “**Because my big**,** mean dog is sleeping**.”**
>
> **“**Oh my**!”** said Angela**. “**Does he bite**?”**
>
> José smiled and said**, “**Only when he's awake**!”**

The punctuation is missing in the story below. Add the missing punctuation as you read. You will need to add periods, commas, quotation marks, exclamation marks, and one question mark.

The batter swung with all his might Crack The ball sailed high toward Billy in right field

Catch it Catch it called Mike at first base Do you have it

Billy looked up He yelled Mine The ball slowly fell down toward his glove Billy watched it second by second Finally it landed He felt its weight in the pocket of his glove

Out yelled the umpire

The team went wild We won We won they shouted

The coach ran onto the baseball diamond He smiled at each player and said Great game You worked hard to win this one Way to go

What a perfect way to end a perfect season

Write to Know Trivia

When Babe Ruth hit sixty home runs in 1927, he hit fourteen percent of all home runs in his league that year. For a player to hit fourteen percent of all home runs today, he would have to hit over 300 home runs in one season!

0-7696-4026-5 *Writing*

Name ______________________________ Date ______________

Your Turn! A Fable, Part 1

Now it's your turn to write a modern fable. Choose one of the morals below. Create a story that represents the moral you have chosen. Your characters can be animals that act like people, as in Aesop's fables. Find examples of fables in the library or on the Internet to help you with format and style. Plan the story carefully. Use the story map below to organize your fable.

Be careful what you wish for.	Little friends may be great friends.
It is better to work smart than to work hard.	Look before you leap.
Looks can be deceiving.	Honesty is the best policy.

Moral: ______________________________

STORY MAP

Characters (description)

Setting (time and place)

Problem

Events

Solution

Name ______________________________ Date ______________

Write to Narrate

Your Turn! A Fable, Part 2

Use your story map to write a rough draft of your fable. Remember to use punctuation correctly. Be sure to use dialogue to show what your characters are saying.

Write to Know Trivia

Aesop's fables were first shared as oral stories over two thousand years ago. They were passed down from one generation to the next only through storytelling for several hundred years.

0-7696-4026-5 *Writing*

Name ______________________ Date ______________

Tools to Use: Word Processing Tips

After writing a rough draft of your fable, use a word processing program to type it into a computer. You can work on your writing onscreen until you have a final draft.

TIPS BEFORE YOU TYPE—Setting Up Your Document

☞ Look under **File** in your menu bar. Select **Page Setup**. Under the **Margins** tab, you can set your page margins. Margin settings are usually set for 1" top and bottom and for 1.25" left and right.

☞ Now, choose your **Font**. Look under **Format**. Select **Font**. Choose a font that is easy to read. Fancy fonts are not good for formal writing. The most common fonts are Times, Times New Roman, Helvetica, and Arial. Choose the size next. Most papers are written at size 12.

☞ Set your spacing for double spacing. Look under **Format**. Select **Paragraph**. Spacing choices usually include single, 1.5 lines, and double. Select **double**. This creates a line of space between each typed line.

☞ You will need to indent the first line of each new paragraph. There are two ways to do this. The first way is to hit the **Tab** key as you begin a new paragraph. Tab will automatically indent, usually half an inch. Or, you may look under **Format** and select **Paragraph**. You will find the words **Indentation** and **Special**. Choose **First line** under **Special**. Then type in the number 0.5 beside **Left** for a left-side indentation. Click on **OK**. Your document is now set!

TIPS AFTER YOU TYPE—Checking and Correcting Your Document

☞ After you have typed your story, you can check your work. Begin by doing a spelling and grammar check. Look on your menu bar under **Tools**. Select **Spelling and Grammar**. Click **OK**. Correct any misspellings or incorrect grammar you may have had.

☞ Review your work onscreen. If it looks ready, then print the story. You may review the final pages by selecting **Print Preview** under **File** on the menu bar. If it looks correct, print the pages. Under **File**, select **Print**. If you want more than one copy, enter the number of copies you want. Click **OK**. Your final draft should print out perfectly!

Name ______________________ Date ______________

Revise for the Prize

Use the tips on page 48 to create a printed copy of your fable. Use this checklist to make sure your final copy is as perfect as possible.

- ❑ All words are spelled correctly.
- ❑ All sentences are complete.
- ❑ All paragraphs are indented.
- ❑ All punctuation is used correctly.
- ❑ My story has a beginning, middle, and end.
- ❑ My story uses dialogue correctly.
- ❑ My story has a problem that is solved in the end.
- ❑ My fable clearly represents the moral I chose.

Draw a picture that illustrates the fable you have written. When you are finished, display this with your fable. Share your finished products with your classmates. Then, add them to your Personal Portfolio.

Write to Know Trivia

On April Fool's Day, 1976, Steve Wozniak and Steve Jobs released the Apple I computer and started Apple Computers.

0-7696-4026-5 *Writing*

Name ______________________________ Date ____________________

Your Turn! Make a Book

To **publish** something means to make it public. Anyone can be a published writer! One way you can publish your writing is to create a book. For this, you will need:

- several pieces of blank 8 ½" x 11" paper
- a pair of scissors
- tape or glue and a stapler
- a new copy of the final draft of your writing piece
- pencils, markers, or other drawing tools

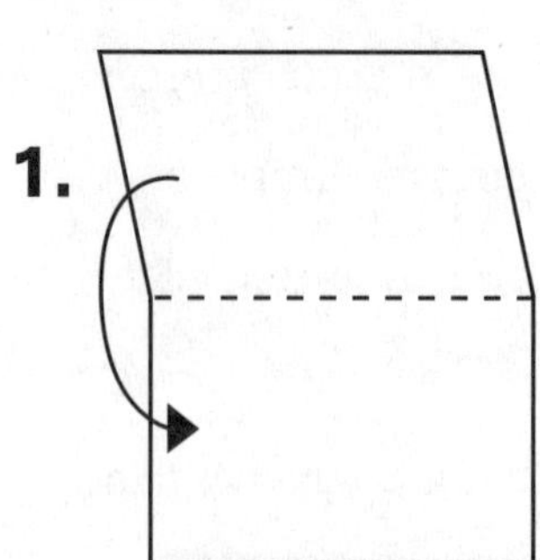

Directions for making your book:

1. Fold each sheet of paper in half, top to bottom.
2. Fold each sheet again, side to side.
3. Insert one folded sheet in the center of the other as shown. Use as many sheets as you want for the book, depending on the length of your writing piece.
4. Staple the sheets along the center fold.
5. You now have a multi-page book.
6. Draw a cover for your book. Create a title for your piece.
7. Cut and paste the words from your printed work onto the book pages. Think about where you want to start and stop your words on a page. Be sure that the story moves clearly from page to page.
8. Illustrate your book.
9. Share your book with others who have made theirs, too.
10. Share your book with a young child.

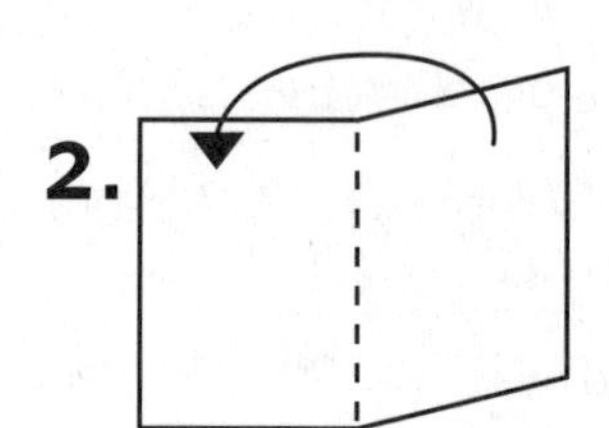

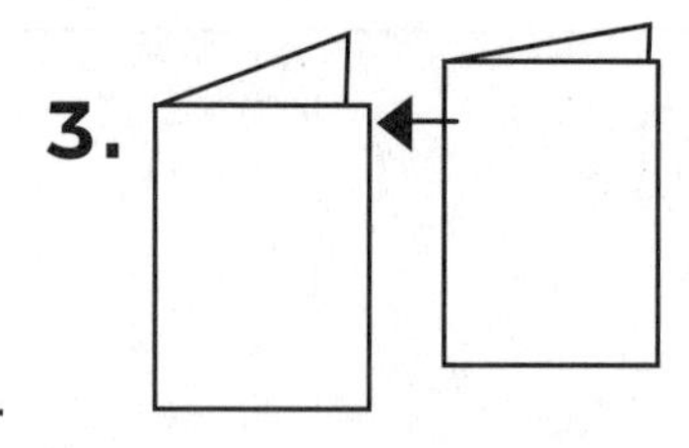

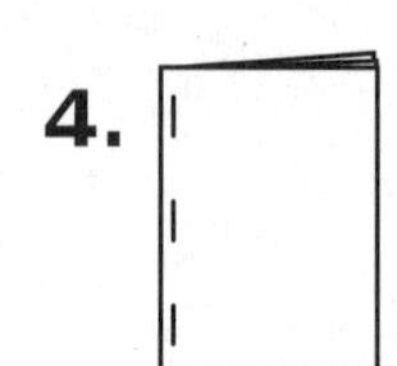

Name ______________________ Date ______________

Write to Persuade

The Real Deal

Read the following writing sample. Think about where you might find a paragraph like this.

UNFAIR FERRET LAW

As a citizen of this city, I think the no-ferret law is unfair. Ferrets are small animals that make great pets. They are not wild animals, even though they look a lot like weasels. People have been keeping ferrets as pets for hundreds of years. They are tame and can be trained. Ferrets are not normally dangerous. Each year, many more people are bitten by dogs than by ferrets. Ferrets are not the best pets for young children, but they are very good companions for the rest of us. Ferrets get vaccines, just like cats and dogs. They do not spread disease or carry rabies. It is unfair to keep responsible people from owning these lovable animals. I think the law should be changed right away.

1. What is the issue the writer is concerned about?

____

____

2. What is this writer's purpose for writing to the newspaper?

____

____

3. What reasons probably led to ferrets being outlawed in this city?

____

____

____

4. How does the writer deal with those reasons?

____

____

____

____

5. After reading this, what do you think about the issue? ______________________

____

In your opinion, how effective was this persuasive writing? Circle your score.

1 2 3 4 5 6 7 8 9 10

Not effective — Somewhat effective — Very effective

Write to Know Trivia

Female ferrets are called jills, and males are hobs. Baby ferrets are called kits.

Name ______________________________ Date ______________

Words to the Wise: Fact or Opinion

Everyone has opinions about things. Persuasive writing begins with an opinion. An **opinion** is a personal view. It cannot be proven. The writer's purpose is to influence others to think like he or she does. To convince the reader, a writer must use facts to support his opinion. A **fact** is something that can be proven. It is known to be true.

The milk tastes bad. OPINION

The milk is two weeks older than its expiration date. FACT

Read each sentence below. If the sentence is an opinion, write *opinion* on the line. If it is a fact, write *fact* and then list two places where you might find proof that the sentence is true.

1. Michael Jordan is the best basketball player ever to play the sport.

   ____

2. In 1933, the character Mickey Mouse received more than 800,000 fan letters.

   ____

3. Sharks are strange-looking creatures of the sea.

   ____

4. Pluto is usually the farthest planet from the sun and has one satellite.

   ____

5. A human being loses an average of 40 to 100 strands of hair a day.

   ____

Read the three quotes below. Choose one. Write what you think it means and whether you agree with it or not.

"You're entitled to your own opinions.
You are not entitled to your own facts." (Anonymous)

"Every story has three sides.
Yours, mine, and the facts." (Foster M. Russell)

"Every man has a right to be wrong in his opinions.
But no man has a right to be wrong in his facts." (Bernard M. Baruch)

____

____

____

0-7696-4026-5 *Writing*

Name ______________________ Date ______________

Tools to Use: Fact-Checking Tips

Facts can be proven. But where do you find the proof you need? Here are some common sources of information.

Encyclopedias—general information and basic facts
Almanacs—specific information and facts about space and the earth's atmosphere
Experts—people who have studied or know about certain subjects
Magazines—general information and facts about specific subjects and current events
Newspapers—current events

About the World Wide Web

Be careful when using sources on the Internet. Anyone can create a Web site. It is important to find out who created the site. Try to figure out how reliable the information is. The Web site address (URL or domain name) ends with letters that can help you.

.com or .net = a company, organization, or person. These are the most common endings.
.gov = government agency
.edu = educational place, such as a college or school district
.org = a nonprofit group or organization

When researching topics on the Internet, ask yourself the following questions:

Who is the author? Can you tell by the Web address?
Is the information reliable?
Is the individual or group that created the site trustworthy?
Is there a way to contact the creator or Webmaster of the Web site?
(like an email or postal address, telephone number, etc.)

Read each opinion below. Where could you find facts to support the opinion?

1. Teenagers spend a lot of money. ______________________

____

2. Cats make the best pets. ______________________

____

3. Tanning salons should be banned. ______________________

____

4. Fast-food restaurants should serve only healthy food. ______________________

____

Name ______________________________ Date ______________

Words to the Wise: General or Specific

Read the following pair of persuasive paragraphs.

Today's kids are not very healthy. Most kids do not get enough exercise. Their diets are not balanced. Kids eat a lot of snack food rather than fruits, vegetables, and grains. This is true at home and at school. Many kids become overweight at very young ages. Some diseases that used to appear mainly in adults are now appearing in children. Unless kids change their habits, they will grow to be unhealthy adults.

Today's kids are not as healthy as they could be. National health reports show that young children are not getting enough physical exercise, even in their school programs. The National Center for Health Statistics states that more than eight million U.S. kids, ages 6 to 19, are overweight. The director of this center says that obesity and lack of physical fitness in our young children may cause diabetes, heart disease, and other serious health problems later in life.

Are both paragraphs persuasive? ______________________________

Which paragraph is more effective? Why? ______________________________

____

Read each general statement below. On the line provided, write a specific statement that is more effective. You may need to do some research.

1. Barry Bonds is the best baseball player. ______________________________

   ____

2. Niagara Falls is a popular vacation site. ______________________________

   ____

3. A one-dollar bill doesn't last very long in circulation. ______________________________

   ____

4. California has had a lot of earthquakes. ______________________________

   ____

Name ______________________________ Date ______________

Your Turn! Write an Ad

You can have the whitest teeth ever! Quick, easy, and proven! Dentists agree that XYZ Tooth Polish works—and it's good for you. Cavities stay away when you use XYZ! And that is something to smile about! On sale now at a store near you.

What is this ad doing to persuade a person to buy the product?

____

____

____

____

____

____

____

Choose one of the following items and write your own ad. Use strong words to lure the reader to want your product. Make it as persuasive as you can. Illustrate it, too. Design and art are part of the effectiveness of an ad!

Anti-Stink Athletic Socks
"Bad Hair Day" Hair Spray
Plant Pots with Super-Dirt
Free Puppies (or Kittens)

____

____

____

____

____

____

____

____

Write to Know Trivia

The word Velcro® came from the French words "velour," meaning velvet, and "crochet," meaning hook.

Name ______________________________ Date ______________

Your Turn! Persuasive Paragraph

Do you have an opinion about a topic that is important to you? Write a list of issues that matter to you.

____

____

____

About which of these issues do you feel the strongest? ______________________________

Write a topic sentence that shows your opinion.

____

What facts and details can you use to support your opinion?

____

____

____

____

____

Write a concluding sentence that persuades your reader.

____

Put all of these together to write your own persuasive paragraph.

____

____

____

____

____

____

0-7696-4026-5 *Writing*

Name ______________________ Date ______________

Revise for the Prize

Think of ways to make your persuasive paragraph more effective. Answer these questions.

1. Which words in my rough draft can I make stronger? ______________

2. What stronger words can I use instead? ______________

3. Is my topic sentence as clear as it can be? Does it truly show my opinion? ______________

4. Which facts and details best support my opinion? ______________

5. What other facts could I use that are more effective? ______________

6. What sources should I use to make sure my facts are correct? ______________

7. Is my concluding sentence as strong as it can be? If not, what is a stronger conclusion? ______________

8. Have I corrected all of my spelling, punctuation, and capitalization errors? ______________

Revise your rough draft until your paragraph is as perfect as you can make it. Write a final draft in ink or print it using a computer. Put it in your Personal Portfolio. Congratulations! You have just written a persuasive paragraph.

Write to Know Trivia

Even big newspapers make mistakes. In 1948, the *Chicago Tribune* declared, "Thomas Dewey Defeats Truman." In fact, Harry Truman had won the presidential election.

Name ______________________ Date ______________

Read and Write: Literature Response

Read the following historic letter. It was written in October of 1860 to a man who was running for President of the United States. His name was Abraham Lincoln. At the time, Lincoln did not have a beard.

Dear Sir:

My father has just [come] home from the fair and brought home your picture and Mr. Hamlin's. I am a little girl only 11 years old, but want you to be President of the United States very much. So I hope you won't think me very bold to write to such a great man as you are. Have you any little girls about as large as I am? If so, give them my love and tell her to write to me if you cannot answer this letter. I have got four brothers and part of them will vote for you anyway. If you let your whiskers grow, I will try and get the rest of them to vote for you. You would look a great deal better for your face is so thin. All the ladies like whiskers, and they would tease their husbands to vote for you, and then you would be President. My father is going to vote for you, and if I was a man I would vote for you too. But I will try to get everyone to vote for you that I can. I think that rail fence around your picture makes it look very pretty. When you direct your letter, direct to Grace Bedell, Westfield, Chatauqua County, New York. I must not write anymore. Answer this letter right off. Good-bye.*

Grace Bedell

*letter adapted for readability

1. What is Grace trying to persuade Abe Lincoln to do? ______________________

____

2. Why does she want him to do this? ______________________

____

3. Do some research. Find out if her letter was effective or not. Did she get an answer to her letter? Write your findings below. ______________________

____

____

____

Name ______________________________ Date ______________

Your Turn! A Persuasive Letter

Grace Bedell was only eleven years old when she wrote to Abraham Lincoln. You can write to elected officials, too. What would you like to persuade the President to do?

Write a persuasive letter to the President. Explain what you would like him to do. Then give reasons for your request. Support your request with details and facts.

Dear Mr. President,

____

____

____

____

____

____

____

____

____

____

____

____

Sincerely yours,

____

Write to Know Trivia

President Ronald Reagan was quite a letter writer! In his lifetime, President Reagan may have written more than 10,000 letters!

0-7696-4026-5 *Writing*

Name ________________________________ Date ________________

Your Turn! One Side of the Coin

A good writer can see both sides of an issue. You will write two essays on one topic. The first essay will be in favor of something. The next one will be against the same topic. Choose one of the topics below. Write a short essay made up of at least three paragraphs: introductory, supporting, and concluding.

Children under the age of 16 should have to wear helmets while skateboarding, bike riding, skating, and skiing.

School should be in session all year long.

Field trips are a good use of money and time.

Beginning at age 14, a person should be able to have a credit card.

My Topic ________________________________

Notes for Introductory Paragraph

____

____

Notes for Supporting Paragraph(s)

____

____

____

____

____

Notes for Concluding Paragraph

____

____

____

Consider these points to make sure you have fully supported your argument.

Have someone with the opposite viewpoint read over your supporting details.

Use specific, strong words to make your sentences more powerful.

Try to find facts to support your side. It's hard to argue against facts.

0-7696-4026-5 *Writing*

Name ______________________ Date ______________

Your Turn! The Other Side of the Coin

Now think about the "other side" of the issue. Choose the statement below that has the opposite viewpoint of the one you chose on page 60. Write a short essay made up of at least three paragraphs: introductory, supporting, and concluding.

Children should not have to wear helmets while skateboarding, bike riding, skating, and skiing.

School should not be in session during the summer.

Field trips are not a good use of money and time.

A person under the age of 18 should not be able to have a credit card.

My Topic ______________________

Notes for Introductory Paragraph

____

____

Notes for Supporting Paragraph(s)

____

____

____

____

____

Notes for Concluding Paragraph

____

____

____

Why is it good to be able to write from both sides of an issue?

____

____

Name ________________________________ Date ________________

Revise for the Prize

Choose the essay you most agree with from pages 60 and 61. Answer the following questions before you revise your essay.

1. Who is a possible audience for this essay? Who are you trying to convince or persuade?

____

____

2. How would your essay be different if your audience was a young person?

____

____

____

3. How would your essay be different if your audience was an adult?

____

____

____

4. What sources could you use to find facts to support your opinion?

____

____

5. Who would be a good person to "try it out" on? ____________________

6. Let that person read your essay. What improvements does he or she think you could make?

____

____

Finishing Touches

Make any changes that will improve your essay. Correct any mistakes. Write the essay in ink or print it out using a computer. Place it in your Personal Portfolio. Congratulations! You have written a persuasive essay.

Name ______________________________ Date ______________

Read and Write: Literature Response

The Blind Men and the Elephant

by John Godfrey Saxe (1816–1887)

It was six men of Indostan
To learning much inclined,
Who went to see the Elephant
(Though all of them were blind),
That each by observation
Might satisfy his mind.

The First approached the Elephant,
And happening to fall
Against his broad and sturdy side,
At once began to bawl:
"God bless me! but the Elephant
Is very like a wall!"

The Second, feeling of the tusk,
Cried, "Ho! what have we here
So very round and smooth and sharp?
To me 'tis mighty clear
This wonder of an Elephant
Is very like a spear!"

The Third approached the animal,
And happening to take
The squirming trunk within his hands,
Thus boldly up he spake:
"I see," quoth he, "the Elephant
Is very like a snake!"

The Fourth reached out an eager hand,
And felt about the knee.
"What most this wondrous beast is like
Is mighty plain," quoth he;
"'Tis clear enough the Elephant
Is very like a tree!"

The Fifth, who chanced to touch the ear,
Said: "E'en the blindest man
Can tell what this resembles most;
Deny the fact who can
This marvel of an Elephant
Is very like a fan!"

The Sixth no sooner had begun
About the beast to grope,
Than, seizing on the swinging tail
That fell within his scope
"I see," quoth he, "the Elephant
Is very like a rope!"

And so these men of Indostan
Disputed loud and long,
Each in his own opinion
Exceeding stiff and strong,
Though each was partly in the right,
And all were in the wrong!

Answer the following questions.

1. How did each man "research" the animal? ______________________________

2. Each man was wrong about what the animal was. But each man had one part right. How might they have come up with an accurate report?

____

____

3. What lesson can you learn from this story that applies to doing a research report?

____

____

0-7696-4026-5 *Writing*

Name ______________________ Date ______________

Tools to Use: A Topic Graphic Organizer

The first step in writing a report is making a plan. A good writer organizes what he or she already knows, needs to know, and makes a list of questions to research. This will help you focus your topic. Look at the example below. What kind of facts and questions could you add to this chart?

My Topic: Dogs' Health

Facts I Know	How I Know These Facts	Questions I Need to Research	Sources of Information
Puppies need good food to grow strong.	I read ads about dog food and the information on food packaging.	What exactly do puppies need to eat and how much? What happens if they do not eat well?	• My vet • Books about pet care • Articles from pet care magazines
Dogs need shots against diseases and yearly visits to the vet.	My vet recommends certain foods and shots.	What diseases can puppies get? When and how should they get their shots?	• My vet • Books about pet care
Puppies need to chew things.	My puppy chews on lots of things because he's getting new teeth.	What makes a puppy want to chew? What is a safe thing for them to chew on?	• Books about pet care • Web sites from pet organizations
Puppies need to be trained to be good pets.	My puppy makes lots of mistakes, especially on the floor.	How can puppies be potty trained? What kinds of training tips work well for dogs?	• Articles from pet care magazines • My vet • Books

Name ______________________________ Date ______________

Words to the Wise: Focusing Your Topic

Beginning a research report can feel a little overwhelming. Sometimes the topic you are interested in is either too narrow or two broad. Spend some time thinking about your topic. Ask yourself these questions:

- What topic did you choose?
- Why did you choose this topic?
- What do you find interesting about this topic?
- What would you like to learn more about regarding this topic?
- Are there any special issues or problems with this topic?
- What do you want your readers to remember most about your paper?
- What is the main message for your topic?

Questions will help you focus on your topic. Answers to those questions will help you focus and narrow your direction before you begin researching.

Almost all assignments, no matter how complicated, can be reduced to a single question. Your first step, then, is to reduce the assignment into a specific question. Even if your assignment doesn't ask a specific question, your focused topic still needs to answer a question about the issue you'd like to explore. In this situation, your job is to figure out what question you'd like to write about.

The focused topic at the start of your work may need to change as you find new information. That's okay. It is part of the research. You will need to have a final focused topic before you begin your rough draft, though.

Here are some examples of focused topics:

Identical twins often have special connections with each other.
Babe Didrickson was one of the greatest female athletes of all time.
Lowering the legal driving age has many positive and negative effects.
An adult dog will have fewer health problems if he is well cared for as a puppy.

These statements will appear somewhere in each one's introductory paragraph. They are the main ideas that drive those papers. When the reader sits down with the report, you want him or her to be able to easily identify your focused topic.

Name ______________________________ Date ______________

Words to the Wise: Sources

You can find information in many ways. You can talk to people—experts, professionals, or specialists. You can go to places—libraries, zoos, museums, or historic sites. You can also look at things—books, magazines, newspapers, or the Internet. Each of these has information, but each has a different kind of information. It's like the elephant, with many parts that are all important to the whole.

Here is an example of an organizer for identifying sources. The focused topic for this student's report is: An adult dog will have fewer health problems if he is well cared for as a puppy.

Places to Find Sources		
People	**Places**	**Things**
dog breeders	veterinarian's office	magazines about dogs
veterinarians	pet shop	books about raising puppies
Humane Society	dog owners	Internet Web sites by dog professionals
	dog food company	reports from veterinary journals

Brainstorm a list of people, places, and things you can use as sources for your report. Don't be afraid to add people and places. People will want to help you!

People	Places	Things

Name ______________________________ Date ____________________

Your Turn! Taking Notes, Part 1

Once you have found sources for your report, it is time to pull information from them. Begin by organizing your sources. You will need several 3" x 5" index cards. Each source will get its own card. This will be very helpful when you write the bibliography at the end of your report. A **bibliography** is a list of the sources you used for your report's facts.

On each card, collect the following information about each of your sources. Use these formats and examples when you write your cards. (Note: Examples are in MLA format.)

Books Author (last name, first name). Title of book underlined. City where book was published: Name of publisher, Date of publication.

Ex.: McLennan, Bardi. <u>Puppy Care and Training</u>. New York: Howell Book House, 1996.

Articles Author of article (last name, first name). Name of article in quotation marks. Name of publication underlined. Date the article was published, page numbers of article.

Ex.: Ritzovich, Uri. "Pups and Training." <u>Dog Fancy Magazine</u>. June 2003, pp. 42–48.

Interviews Name of person interviewed (last name, first name). Personal interview. Date of interview (day month year).

Ex.: Voss, Dr. John, DVM. Personal interview. 23 September 2004.
Lopez, Guadalupe. Telephone interview. 27 September 2004.

Online sources Name of article (if given) in quotation marks. [online article]. Web site's organization. <URL network address>.

Ex.: "Training Puppies." [online article]. American Kennel Club. <http://www.akc.org>.

When you have a card for every source, put the cards in alphabetical order by either the author's last name or the article name (if no author is given). Give each source its own number. This will be helpful when you begin taking notes from each source.

Name ______________________________ Date ______________

Your Turn! Taking Notes, Part 2

Now that you have your sources, it's time to get the facts. Here are tips for taking terrific notes.

1. Get some index cards to write notes on. They can be 3" x 5" or 4" x 6".
2. Begin by reading one page at a time, and get all the information you can from it.
3. Number each note card with the same number you assigned the source. This will help keep your note cards together. For example, if you are working with a book that is number 3 in your list of sources, then make sure all the note cards you write from this book have a number 3 written on the corner of those cards.
4. Write down the most important facts and details that will help you build your report. Include the page number where you find each piece of information. This will help you find them again if you need to.
5. When you write exact words from a source, place quotation marks around it. Write down who said the quote so you can give credit to that person in your report.
6. Most of your notes should not be exact quotes, though. You should **paraphrase**, or use your own words. This prevents people from stealing others' thoughts.
7. Look up unfamiliar words and add the definitions to your cards.
8. Put a general heading on each card to show the main idea of the card. This will help you divide your notes into the body paragraphs later.
9. For an interview, ask to tape record the person's answers and then take your notes from the tape.
10. If you find information you aren't sure you'll use, write it anyway. It is better to have too many notes than not enough!

Messonier, Shawn. 8 Weeks to a Healthy Dog. Emmaus, Chicago: Rodale Publishing, 2003. pp. 5-12 — 3

Triangle of Dog Care — 3
Three important parts to dog health: you, your vet, the dog
Owner is key. Knows the dog best. Can tell when things are wrong.
Choose best vet. Helpful and answers questions.
(page 5)

Name ______________________________ Date ________________

Tools to Use: The Outline

Your note cards will help you create an outline. Look through your cards. Make stacks according to the headings on the cards. Then put the cards in an order that makes sense. Order can be based on time (first, second, etc.). It can be based on direction (top to bottom, left to right, small to large, etc.). Your outline can also be organized from general to specific information. It all depends on your focused topic. You are building a report. Build it clearly. Here is an example. Pay attention to the form.

Focused topic: An adult dog will have fewer health problems if he is well cared for as a puppy.

I. People are important
  A. Breeder
  B. Owner
  C. Veterinarian
  D. Groomer
  E. Others

II. Healthy puppies need these
  A. Good food
  B. Lots of rest
  C. Exercise
  D. Shots and medicine
  E. Training

III. How healthy puppies grow
  A. Physical growth
  B. Behavior growth
  C. Dealing with illness and injury
  D. Lifelong care

Conclusion: Tender-loving care can help a playful puppy grow into a perfect pet.

Write It!

Follow the form above to write your own outline. Use your own paper. Make changes as you do more research. If you find that things are missing that you would like to include, go back and add to your research.

0-7696-4026-5 *Writing*

Name ______________________________ Date ______________

Your Turn! The Rough Draft

The next part of writing a report is like making a banana split. First, you need to get all of your materials together.

Focused topic	This is like the name of the dessert you are making.
Sources	This is like your list of ingredients.
Note cards	These are like your ingredients.
Outline	This is like your list of directions.
Final report	This is the dessert put together at last and ready to be eaten.

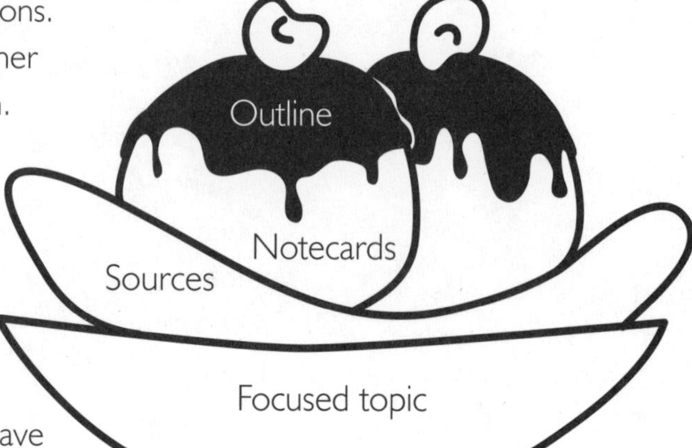

____

Use the following tips for writing your rough draft.

1. Revise your outline if needed.
2. Write the introduction. The introduction should have two parts. The first should get the reader's attention. The second part should state your focused topic. You might want to give your reasons for choosing this topic.
3. Next, write the body of the report. These are the paragraphs that support your focused topic. Look at your outline and notes together. Follow the outline to create paragraphs for each section.
4. Make sure each paragraph has a topic sentence. Build each paragraph as usual—with supporting details and examples. The concluding sentence should wrap up the main idea for that paragraph. Use transitions between paragraphs to give the report direction.
5. Write the report in your own words. If you use an exact quote, remember to give credit to the source.
6. Write a conclusion that summarizes your report and leaves your readers with a final statement that they will remember.

Your Turn!

Use the tips and examples given to write your rough draft. Remember to keep returning to your focused topic. If you have details that do not quite fit the topic, do not include them in your report. Look for areas that could use some development. You might have to go back and research a little more to strengthen a particular point. It is better to do that now rather than when you are ready for proofreading and revising.

Name ______________________ Date ______________

Words to the Wise: Proofreading Basics

After you have written your rough draft, review it using proofreading marks. Use a colored pen to make your corrections so they are easily seen when you go back to fix them later.

≡ means to capitalize the above letter.

⊙ means to add a period here.

—ℓ means to delete what is crossed out.

∧ means to insert something here. This symbol is called a caret.
(If the inserted material is lengthy, write it in the side margins, and draw an arrow to the caret.)

If you have numerous corrections, rewrite your rough draft before doing the next part of proofreading. Have a classmate or someone else read it. Have that person check for content, spelling, grammar, and organization. When the person has finished, ask your reader the following questions:

- ❑ Did you notice any misspelled words?
- ❑ Are there any problems with grammar or punctuation?
- ❑ Are my sentences complete (no fragments or run-ons)?
- ❑ Is my focused topic clear and organized enough?
- ❑ Does the order of my writing make sense?
- ❑ Does each body paragraph support the focused topic?
- ❑ Are my topic sentences clear and complete?
- ❑ Are my sources reliable and varied?

Listen closely to any suggestions for improvement. This is the perfect chance to fix anything and make your paper the strongest it can be.

0-7696-4026-5 *Writing*

Name ______________________________ Date ____________________

Words to the Wise: Bibliography Basics

After proofreading your rough draft, take a look at your list of sources. Every research paper has a bibliography (or works cited) page. The format of this page must follow certain rules.

Each source will have its own **entry**, like a dictionary does. Each entry is listed in alphabetical order, by the author's last name. If there is no author, it is listed by the first word of the title. The second line of each entry is indented, so it is easier to find a particular listing.

Use your source cards to create a rough draft of your bibliography. Ask your teacher for help if you cannot find the format you need.

Here is an example. (Note: Entries are in MLA format.)

Works Cited

Adamson, Eve. "Pups Can't Live on Love Alone." Dog Fancy Magazine. June 2003, pp. 28–32.

Hopkin, Karen. "Are Dogs Dumb?" Muse Magazine. November/ December 2002, pp. 119–125.

Kilcommons, Brian and Sarah Wilson. Good Owners, Good Dogs. New York: Warner Books, 1999.

King, Dr. Samuel, DVM. Personal interview. 21 September 2004.

McLennan, Bardi. Puppy Care and Training. New York: Howell Book House, 1996.

Messonier, Shawn. 8 Weeks to a Healthy Dog. Chicago: Rodale Publishing, 2003.

"Pups Grow Up." Animal Planet. 28 July 2004.

"Training Puppies." [online article]. American Kennel Club. <http://www.akc.org>.

0-7696-4026-5 *Writing*

Name ______________________ Date ______________

Tools to Use: Formatting the Final Report, Part 1

The best presentation for a research paper can be made on a computer. Your earlier drafts may have been handwritten. Now it is time to type your final draft into a word processing program. See the samples below of finished pages. Follow the examples when formatting your final draft.

(the cover page)

The Importance of Good Puppy Care
by Samantha Adams
Mr. Williams
Language Arts 6-A
October 28, 2004

(the outline)

Focused topic: An adult dog will have fewer health problems if he is well cared for as a puppy.

I. People are important
   A. Breeder
   B. Owner
   C. Veterinarian
   D. Groomer
   E. Others

II. Healthy puppies need these
   A. Good food
   B. Lots of rest
   C. Exercise
   D. Shots and medicine
   E. Training

III. How healthy puppies grow
   A. Physical growth
   B. Behavior growth
   C. Dealing with illness and injury
   D. Lifelong care

Conclusion: Tender-loving care can help a playful puppy grow into a perfect pet.

Name ____________________ Date ____________

Tools to Use: Formatting the Final Report, Part 2

(the report)

The Importance of Good Puppy Care

Healthy puppies do not just happen. It takes many people, practices, and products to make little puppies grow into healthy dogs. Careful planning and commitment for dog care is not easy. Raising a pet is a big responsibility. Everyone who owns a pet should be willing to work hard to keep his or her pet healthy.

Key people, such as veterinarians, breeders, and owners, must work together for the sake of the dog. They must be able to communicate for t dog, because the dog cannot. Veterinarians understand how a dog's body works. They c medicine when needed. Sometimes other procedures must take place. Surgeries, such spaying or neutering, are best for a dog's hea

1

(the bibliography)

Works Cited

Adamson, Eve. "Pups Can't Live on Love Alone." Dog Fancy Magazine. June 2003, pp. 28–32.

"Healthy Dog." [online article]. American Kennel Club. <http://www.akc.org>.

Hopkin, Karen. "Are Dogs Dumb?" Muse Magazine. November/December 2002, pp. 119–125.

Kilcommons, Brian and Sarah Wilson. Good Owners, Good Dogs. New York: Warner Books, 1999.

Messonier, Shawn. 8 Weeks to a Healthy Dog. Chicago: Rodale Publishing, 2003.

King, Dr. Samuel, DVM. Personal interview. 21 September 2004.

"Pups Grow Up." Animal Planet. 28 July 2004.

Print your final draft. Place it in a report cover. After your teacher grades it, place it in your Personal Portfolio. Congratulations! You have written a top-notch research report!

0-7696-4026-5 *Writing*

Name ______________________________ Date ____________________

Assessment

Writing a Paragraph

Choose one of the following topics on which to write a paragraph. Remember all you have learned about writing a good paragraph. Your paragraph may be descriptive, expository, narrative, or persuasive. Choose the type that best fits your topic. It's your turn to show what you know!

My Favorite Movie	How to Make My Favorite Meal
Lost in the Forest	Being the Youngest Child
My Hero	My Dream Vacation Spot
Qualities of a Best Friend	The Greatest Invention of All Time

0-7696-4026-5 *Writing*

Name ______________________________ Date ________________

Assessment

Writing a Persuasive Letter

Choose a business in your community where you would like to work. Write a persuasive letter to the owner or manager. Tell him or her why you would like to work there and why you would be a good employee. Use details to persuade the reader to respond to your letter. Follow the correct form for writing a letter.

Name ______________________________ Date ____________________

Assessment

Writing an Expository Essay

Write a five-paragraph expository essay on one of the topics below. You will need to narrow and organize your topic before writing.

A Mystery in History	A Famous Animal	The American Flag
A Natural Disaster	The Wild West	A Planet in Our Solar System

Use the graphic organizer below to plan your essay. Then write a rough draft on another piece of paper. After revising, write a final draft in ink or print it out using a computer and printer.

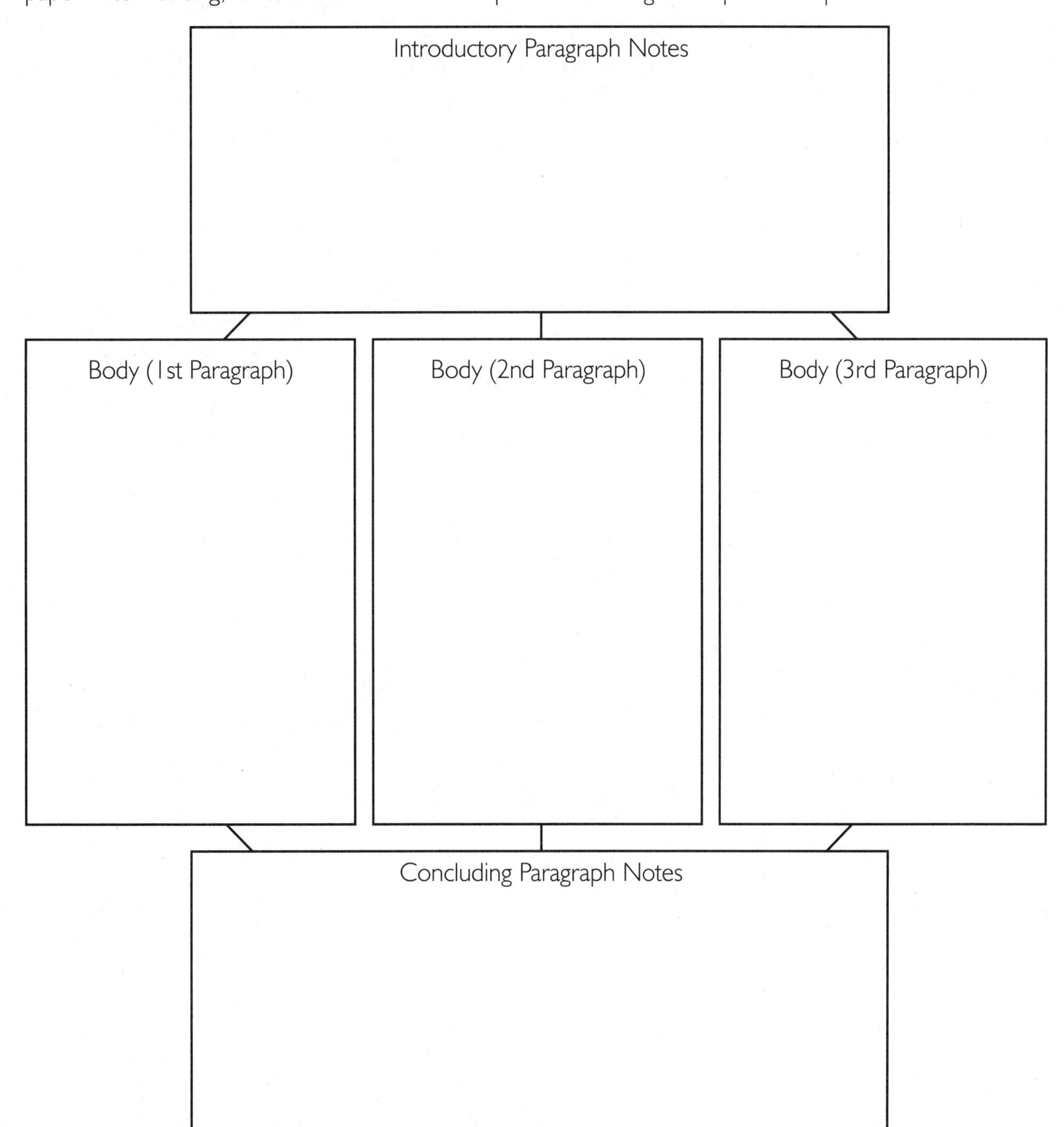

Answer Key

Words to the Wise: Complete Sentences......7

1. sentence
2. fragment; They are big bears that stand up to seven feet tall.
3. fragment; Grizzly bears eat mostly berries, nuts, and roots.
4. sentence
5. run-on; Grizzled means "silver-tipped." It happens as bears get older.

Tools to Use: Punctuation 8

Alaska is home to many kinds of bears. (underline)

Can you name them? (circle)

Black bears, brown bears, and polar bears all live in Alaska. (underline)

It's no wonder Alaska is known as Bear Country! (or period) (underline or box)

Did you know that the brown bear is the largest land mammal in the world? (circle)

Well, it is! (or period) (box or underline)

The adult male can weigh up to 1,500 pounds. (underline)

That is amazing! (box)

Don't get in its way. (or !) (underline or box)

A mother bear will fight to defend her cubs. (underline)

If you ever see a mother brown bear, you should do one thing. (underline)

Leave her alone! (box)

Revise for the Prize.....10

1. Each baseball season, fans eat about 25 million hot dogs. Adults like mustard best.
2. Kids like ketchup best on their hot dogs.
3. Kids like foot-long hot dogs, too.
4. There are many stories about the beginning of the hot dog.
5. Is it really a sausage?
6. Hot dogs became baseball's favorite food in 1893. They were inexpensive and easy to eat.
7. Students from Yale University bought their hot dogs from "dog wagons."
8. The average person takes only about six bites to eat one hot dog.

Focus on Form: Paragraph Structure....11

Paragraph on left:

Skateboarding surfaces make a difference.

Paragraph on right:

Wearing correct skateboarding gear is important.

Concluding sentences will vary for both paragraphs.

The Real Deal 15

The paragraph is explaining allergies caused by poison ivy, poison oak, and poison sumac.

Topic sentence underlined: Some of the itchiest allergies come from three plants.

Facts: Answers will vary. There are several choices in the paragraph.

Concluding sentence double underlined: Unless you like to itch, stay away from these plants!

Focus on Form: Topic Talk 16

1. B
2. E
3. E
4. B
5. E
6. B
7. E

Tools to Use: Where to Find the Facts.........17

Answers will vary. Possible answers:

1. cookbook
2. cookbook, Internet
3. World Record book
4. encyclopedia, Internet
5. health book, Internet, doctor
6. almanac, encyclopedia
7. dictionary, encyclopedia
8. Internet, encyclopedia
9. book about cider, cider maker
10. literature book

Words to the Wise: Transitions.................19

Possible answers, in order:

For example,
When
In the same way,
Another
Therefore,

Tool to Use: Commas ..21

Example explanations:

Teresa is the teacher and she wants to give two things: markers and nametags made of paper.

Teresa is the teacher and she wants to give three things: paper, nametags, and markers.

The writer is talking to someone named Teresa about the teacher who is giving three things: paper, nametags, and markers.

Answer Key

1. When I was little, I had a fish, a hamster, and a rabbit.
2. I won the first race, but then I lost the second.
3. Yes, it was a green, black, and silver insect.
4. Will you bring your brother, Mary?
5. Mr. Andrews, the mayor, came to dinner.

Focus on Form: Letter Format 22

Heading:
455 Elm Street
Boston, MA 02209
June 22, 2004

Inside address:
Mr. Elmer Smith
77 Beachfront Lane, Apt. 3
Myrtle Beach, SC 29578

Salutation: Dear Grandpa,

Body: the paragraph as shown

Closing: Your grandson,

Signature: Benjamin (in cursive and in print)

The Real Deal 27

1. a giraffe
2. It wraps its tongue around leaves, and then chews, swallows, burps it back up, and chews and swallows it again.
3. It is hidden in the trees. It has spots. Its neck is long. It is about 18 feet tall and weighs 4,000 pounds.
4. It gallops on strong hooves.
5. Its spots blend in with the leaves and trees.
6. sight

Words to the Wise: Strong Words..............28

1. weak
2. strong
3. strong
4. weak
5. weak
6. strong
7. weak
8. strong

Possible answers:

9. bird—cardinal
10. car—Mustang
11. walked—strolled
12. small—teeny
13. dark—inky
14. drove—raced
15. said—whispered
16. tall, furry-horned, blinks, long

Tool to Use: Thesaurus 29

Possible answers:

thought—considered
tore—ripped
threw—fired
looking—peering
teacher—professor
need—desire
ancient—long-ago
piece—sample
smiled—grinned
pulled—removed
book—volume
galore—numerous
wonderful—terrific

Tools to Use: Sensory Details 30

Underline: biggest, six feet tall, pointed straight up, giant tan spear wrapped in lettuce leaves, moaned, disgust, Mr. Stinky, five-foot, worst smell, rotting elephant, corpse plant

Senses used: sight, sound, smell

Your Turn! Descriptive Paragraph .31

1. topic
2. support
3. sensory
4. concluding
5. complete
6. indent

Words to the Wise: Connotation................33

1–3. Answers will vary.

4. neutral
5. positive
6. negative
7. negative
8. skeleton-like
9. crude
10. mighty
11. hasty
12. riches

Focus on Form: Haiku..34

Circle: rushing, flows, over, down, sudden, crashing

See: over, down, edge, waterfall

Hear: rushing, crashing

Feel: sudden

Focus on Form: Free Verse...................35

Most used sense: sight

Answers will vary.

Answer Key

The Real Deal 39

Answers will vary.

(Possible answer: making a swimming pool)

Read and Write: Literature Response 42

1. Answers will vary.
2. third-person point of view
3. Dylan, his father, Jody Garner
4. Milford, Iowa; July 4, 2004.
5. how to find the owner of the money
6. Answers will vary.
7. They called the phone number on a piece of paper they found.
8. They did not accept the reward.

Focus on Form: Narrative Paragraph 43

1. an Ant, a Dove, and a bird-catcher
2. along a stream
3. The ant falls in the water, and the bird-catcher tries to catch the dove.
4. The dove drops a leaf in the water for the ant. The ant bit the bird-catcher on the foot to allow the dove to escape.
5. Answers will vary. Possible response: When you do a good thing, sometimes a good thing will happen to you, too.

Words to the Wise: Talk about Punctuation 45

The batter swung with all his might. Crack! The ball sailed high toward Billy in right field.

"Catch it! Catch it!" called Mike at first base. "Do you have it?"

Billy looked up. He yelled, "Mine!" The ball slowly fell down toward his glove. Billy watched it, second by second. Finally, it landed. He felt its weight in the pocket of his glove.

"Out!" yelled the umpire.

The team went wild. "We won! We won!" they shouted.

The coach ran onto the baseball diamond. He smiled at each player and said, "Great game! You worked hard to win this one. Way to go!"

What a perfect way to end a perfect season!

The Real Deal 51

1. The writer of the article feels that a no-ferret law is unfair.
2. The writer would like to see the law changed right away.
3. Students may infer that children have been bitten by ferrets in this city, there is a misunderstanding about the animal in general, and that some pet owners have not been responsible in taking care of their ferrets in the past.
4. The writer lists facts about dangers of other common animals, tries to break myths about ferrets, and argues that responsible pet owners should not be punished.
5. Answers will vary.

Words to the Wise: Fact or Opinion 52

1. opinion
2. fact—encyclopedia or Internet
3. opinion
4. fact—almanac, books, or Internet
5. fact—doctor, books, Internet, or encyclopedia

Words to the Wise: General or Specific......54

Yes, both are persuasive.

The second one is more effective. It uses specific facts to support the opinion.

1–4. Answers will vary.

Your Turn! Write an Ad................55

The ad offers a promise of whiter teeth. It says it is easy, fast, and proven. It tells that dentists approve of the product and that it is even good for you. It says it prevents cavities.

Read and Write: Literature Response 58

1. to grow a beard
2. She thinks he is more likely to be elected.
3. Abe Lincoln did write Grace back and eventually did grow a beard.

Read and Write: Literature Response 63

1. by touching a different part of the animal
2. They could combine their information to decide what it was.
3. Be sure to use several sources for information and do not depend on only one or two.

0-7696-4026-5 *Writing*